EUROPEAN METALS IN NATIVE HANDS

Frontispiece. "Capitaine de la Nation des Illinois." Louis Nicholas (1634–?). *Codex Canadienses,* plate 5. From the collection of the Gilcrease Museum, Tulsa, Oklahoma.

EUROPEAN METALS IN NATIVE HANDS

Rethinking Technological Change 1640–1683

KATHLEEN L. EHRHARDT

THE UNIVERSITY OF ALABAMA PRESS
Tuscaloosa

Typeface: AGaramond

∞

The paper on which this book is printed meets the minimum requirements of American National Standard for Information Science–Permanence of Paper for Printed Library Materials, ANSI Z39.48–1984.

Library of Congress Cataloging-in-Publication Data

Ehrhardt, Kathleen L., 1948–
European metals in native hands : rethinking the dynamics of technological change, 1640–1683 / Kathleen L. Ehrhardt.
p. cm.
Includes bibliographical references and index.
ISBN 0-8173-1440-7 (cloth : alk. paper) — ISBN 0-8173-5146-9 (pbk. : alk. paper)
1. Illinois Indians—First contact with Europeans. 2. Illinois Indians—Industries. 3. Illinois Indians—Commerce. 4. Imports—Mississippi River Valley—History—17th century. 5. Exports—Europe—History—17th century. 6. Indian copperwork—Mississippi River Valley. 7. Copper implements—Europe—History. 8. Technological innovations—Mississippi River Valley. 9. Mississippi river Valley—History—17th century. I. Title.
E99.I2E47 2004
673′.3′08997515—dc22

2004014702

To all the women here and now gone who have helped me along the way.
You inspire me still.
and to
Dr. Sam Nash
Thanks, doll.

The Ilinois, tribes extending toward the South, have five large Villages, of which one has a stretch of three leagues, the cabins being placed lengthwise. They number nearly two thousand souls, and repair to this place from time to time in great numbers, as Merchants, to carry away hatchets and kettles, guns, and other articles that they need. During the sojourn that they make here, we take the opportunity to sow in their hearts the first seeds of the Gospel.

Father Claude Dablon, from St. Esprit Mission, Chequamegon Bay. *Jesuit Relations* of 1669–1671, in Thwaites 1896–1901:54:167.

Contents

Illustrations

Tables

Acknowledgments

This project was born in summer 1994 out of the fortuitous meeting of two seemingly disparate career-shaping forces—my own longstanding intellectual curiosity about how past peoples made and used things, and the opportunity to work at the Iliniwek Village. It came together only through the diligence and generosity of my mentor and advisor at New York University, Rita P. Wright. Vincent Pigott, Karen Blu, Pamela Crabtree, and James A. Brown were pivotal in seeing it through to completion.

Many scholars, administrators, and their respective institutions and agencies have contributed immeasurably to the substance of this work. Archaeometry internships at The Museum Applied Science Center for Archaeology (MASCA) under the direction of Vincent Pigott, Samuel Nash, and Stuart Fleming, and at The Missouri University Research Reactor (MURR) with Michael Glascock and Hector Neff taught me much about the role of materials science in material culture studies. Associates at the Newberry Library tirelessly provided documentary materials. Dr. Charles P. Swann, Bartol Laboratory, University of Delaware, conducted the proton-induced x-ray emission spectrometry (PIXE) analysis. Roger Boyd and his staff at Battle of Athens State Park assisted in artifact processing. Michael O'Brien provided space and equipment (year after year) at the University of Missouri.

Michael Wayman, Ron Hancock, Martha Latta, Penelope Drooker, Charles Rohrbaugh, Robert Hall, Rochelle Lurie, Duane Esarey, James Hunter, Larry Grantham, and Lisa Anselmi commented upon or assisted with important parts of this research.

The bulk of the research behind this book could not have been conducted without generous funding and financial consideration from the Wenner-Gren Foundation for Anthropological Research, the Missouri University Research Reactor (MURR)/National Science Foundation, the U.S. Department of Energy for Reactor Sharing, and MASCA, University of Pennsylvania Museum.

I am deeply indebted to the Upper Mississippi Valley Archaeological Research Foundation, Western Illinois University, and the Missouri Department of Natural Resources for opportunities to work at the Iliniwek Village and for permission to use their collections for analysis. The contributions Larry Grantham, Missouri Division of State Parks, has made to all aspects of this research have been invaluable.

Thanks are extended to Duane Esarey, Frank Abbruzzese, David Westbrook, and Larry Grantham, who put the graphics and maps together.

I also wish to acknowledge the efforts of all the professionals at the University of Alabama Press, Judith Knight, Dawn Hall, Joanna Jacobs, and the production staff who literally made this book happen, and the contributions of the anonymous reviewers, whose comments and suggestions made it a much better product.

Lastly, my appreciation for the ongoing assistance of my fellow scholars, and the unfailing support of my friends and family throughout this process cannot be measured.

EUROPEAN METALS IN NATIVE HANDS

1 / Native Technologies, European Contact, and the Processes and Meanings of Material Change

Archaeologists have always been intrigued by the ways people make and use things. Exploring "continuity and change" in past material, technological, and social systems—that is, asking why *things* change and why they stay the same—has long been at the very heart of inquiry. These questions become even more provocative in situations of early cross-cultural encounter when foreign goods and materials first begin to change hands. For researchers in northeastern North America, recovering native peoples' initial responses to Europeans and their merchandise and identifying the earliest contexts and processes of persistence and change in their material and social systems are a particularly complex and challenging set of problems. As investigators are well aware, within decades of their availability, many European-derived objects and materials appear to have gone quickly from being novelties (desired but not required) to necessities (required) for the native groups who had face-to-face interactions with Europeans and for those who lived farther afield and received them only indirectly through intergroup trade or gift-giving networks. As indigenous demand for and consumption of European goods increased, material repertoires once made up entirely of native manufactures began to transform. In seemingly short order, many "traditional" technologies were altered significantly or abandoned altogether as European objects, materials, and ideas continued to be drawn into native worlds.

Material transformations took place during the same time in which escalating historical, political, economic, and demographic conditions were also exerting disruptive pressures on native social and ideational systems (Rogers

1990:13). Epidemic disease was taking (and in many cases, had already taken [Merrell 1988:96]) its toll on native populations. Demographic upheaval and decline resulted in new kinds of living and group interaction arrangements. Intergroup hostilities escalated as a result of new economic and political alliances forged as a consequence of contact with Europeans and a growing desire not only for their goods but also for their protection against new and old enemies. Subsistence modes, settlement patterns, and daily activities were altered to accommodate European demands for resources. At the same time, missionization and conversion to Christianity effected large-scale changes in native social organization and belief systems.

Trade goods moved quickly into interior North America, often reaching native hands long before Europeans themselves arrived. By the mid-seventeenth century, objects and materials of foreign manufacture had made their way to native peoples of the Western Great Lakes and Upper Mississippi Valley. Goods moved through direct (however sporadic) or indirect contact with French traders, missionaries, and explorers making their way into the interior from the lower St. Lawrence River. Materials also appeared in the interior through various types of exchange activities with other indigenous groups farther to the east. It is also possible that some commodities reaching the midcontinent during this time originated from early Spanish sources in the Southeast or Southwest or from Dutch or English distributors in the East (Drooker 1996a, b). In any case, like their eastern counterparts before them, native peoples in the midcontinent began to accept certain European tools, raw materials, and technologies into their material repertoires. These early responses set in motion important processes of alteration, innovation, and change in their material and technological systems. These are the same patterns of transformation seen in the initial phases of trade farther to the east (Cleland 1992:77).

TECHNOLOGIES IN TRANSITION: GETTING TO THE ROOTS OF CHANGE

In native North American/European contact contexts, indigenous material culture change appears to be alarmingly fast-paced (in many cases), apparently irreversible, yet imminently transformative. For archaeologists, one of the most important measures of early European impact on native cultural systems has traditionally been the extent to which European products were integrated into native material culture repertoires and the increasingly significant role they played in transforming native cultures as contact intensified. Yet, while the final outcome of long-term, sustained native exposure to European manufactured goods has long been known, its beginning phases and earliest contexts are not well documented. Archaeologists have often asserted that initially, assimilation of European objects into native repertoires probably had little effect on native cultural systems, but this assumption largely remains a hypothesis that requires testing at the level of individual cultures (Emerson and Brown 1992:78). Particular cultures and their potentially distinctive culture- and historic-specific responses to the availability of European merchandise, and the technological, social, political, and economic meaning(s) of these responses bear further investigation (Fitzhugh 1985:6). How and why particular groups of native people may have initially sought out and acquired specific types of European-introduced products and materials, what they did with them, in what contexts they used them, and what initial impact they may have had on precontact ways of life have not been adequately researched (but see Bradley 1987; Branstner 1991, 1992; Rogers 1990; Trigger 1985, 1987; White 1983 for notable exceptions).

Elucidating these processes and their historical and cultural implications at the scale of specific cultures is an ambitious undertaking. It involves integrating scant historical evidence from contemporary documentary sources with archaeological data to bring to light the timing, contexts, and particular historical conditions of individual native response to the availability of European items. It also entails exploring the technical, social, and ideological choices involved in acquiring, transforming, and using European objects and materials as they were incorporated into cultural matrices that were also experiencing significant pressures from other sources of disruption such as missionization, epidemic disease, and escalating intergroup warfare.

For researchers in the Western Great Lakes and Upper Mississippi Valley, investigating the nature and timing of material change and attempting to interpret how and in what domains of native life these early transformations first took place have been especially difficult. In many contact situations, material change is thought to have taken place so rapidly as to be barely visible archaeologically. Sites dating to the protohistoric, or "that shadowy period . . . when tangible archaeological evidence indicates European contact well before the advent of written history," are extremely rare (Emerson and Brown 1992:78). Few protohistoric sites or site contexts have been located that contain the earliest evidence of material influence, that is, smatterings of European trade goods of various materials such as iron, copper-base metals, and glass occurring within otherwise indigenous material assemblages. In addition, archaeologists have had an extremely difficult time linking the site locations they do find with specific groups and locales mentioned in contemporary documentary sources. In large part, this has been due to the intense geographic upheaval and dislocation many native peoples of the interior experienced during the mid-seventeenth century as a result of the Iroquois wars (White 1991). Such tangled historical and demographic circumstances have made identifying and tracking individual ethnic groups and the meaning of change in material distribution and use problematic at best.

Yet, it is the material record from these very early contact period sites that remains a most critical and sought after source of data from which to investigate the dynamics of the period and to make interpretations concerning persistence, transition, and/or transformation in selection and use of such raw materials as lithics, clay, osseous materials, glass, and metals (Schortman and Urban 1998:103–104). Archaeologists recognize that these assemblages reflect the availability, acceptance, and integration of certain European weapons, raw materials, and technologies into native material repertoires. As such, they represent the material manifestation of the earliest phases of alteration, innovation, and change in technological systems. Using increasingly sophisticated laboratory, experimental, and contextual approaches, our aim has been to understand more thoroughly the varied roles materials and technologies have played in processes of stability, innovation, interaction, and transformation within and among cultural systems during this crucial, but poorly understood time.

GOALS AND OBJECTIVES OF THIS RESEARCH

In-depth analytical studies aimed at explaining how and in what specific contexts individual native groups, especially those of the Western Great Lakes/Upper Mississippi Valley, responded technologically to the availability of particular types of European trade goods have seldom been undertaken. This work focuses specifically on investigating the responses of one particular native group, the mid-seventeenth-century (late protohistoric/early contact) Illinois, to the availability of one popular type of European-introduced material, copper-base trade metal. The analysis is approached from a technological point of view. Such an approach is not new. In Western Great Lakes archaeology, researchers have frequently attempted to explain processes of native material culture change in contact contexts in technological terms. Traditionally, many of these explanations have focused, either explicitly or implicitly, on the *causative* role that newly available "superior" European goods and technologies played both driving and directing the processes of material and cultural change in native systems. Traditional "acculturation" scenarios featuring native peoples' initial attraction to and ever-increasing but irresistible dependency on technologically superior European goods and materials emphasized their quick and total abandonment of native craft industries, their inevitable cultural dissolution, and their eventual but predictable assimilation to European ways of life (Quimby 1966). Alternatively, "adaptationist" models focused on native need-driven acceptance of recognizably superior European goods to better "fit" them technologically to their environment (Fitting 1976).

Only recently have revisionist approaches arisen wherein archaeologists have begun to view native peoples as independent, active architects of their own histories, exercising a great deal of control over the trajectory, pace, and contexts of transformation within their own cultural systems (Dobres and Hoffman 1994; Rogers 1990; Trigger 1986:258). In terms of their appropriation of European objects and materials, they are no longer assumed to have been "copiers" or "imitators" of European technologies, but to have been creative and innovative in their uses of them (Rubertone 1989:36). Native decisions to accept, reject, redesign, and use European items are based on their own structures of activity, value, and meaning (Branstner 1992; Hosler 1994; Rogers 1990). These realizations have led archaeologists to ask new questions about the complex nature of interactions among native technologies, the availability of new material culture, and continuity and change in technological and social systems in European contact contexts (Branstner 1991, 1992).

This work also challenges the claim that it was the assumed "superior" technological properties of European merchandise alone that initially drew native peoples to foreign weapons, tools, and materials and spurred them to accumulate more and more of them. While it cannot be denied that material change occurred and occurred quickly among some native groups, the notion that material transition was "inevitable," and that the insatiable desire for European goods was uniquely responsible for eroded cultural traditions and crises in native cultural identity is discarded here. Putting aside such deterministic and extremist perspectives, I draw on concepts from the history of technology and the anthropology of technology to resituate technology's role as a variable rather than as a prime mover in the early dynamics and trajectory of native material and social change. As is now well known to archaeologists, European technologies and the items produced from them were initially perceived of and used in quite different ways by native people than their foreign manufacturers originally intended (Bradley 1987; Pfaffenberger 1992:511; N. Thomas 1991:4). These responses are entirely consistent with the notion that all technological activity, that is, the ideas people have about objects and materials, the choices they make, and the actions and processes they engage in when they procure materials and make and use things, is potentially innovative and creative. Importantly, it is imbued with multiple levels of social, economic, and ideological significance within specific cultural systems that are, in turn, acting in and reacting to new and intensifying historical circumstances.

Working within this restructured and resituated view of technology's place in the larger picture of early native/European material engagement, it remains for archaeologists to elucidate how these processes of change unfolded at the microlevel, or at the scale of individual cultures. That is what is new in this book. Herein, I demonstrate that technological responses of specific cultures to the introduction of particular European materials can be revealed through detailed, fine-grained historical and materials analysis, bringing into focus what has been termed the "microenvironments" of change in a culture contact context. Interpreting such "everyday" type responses to new materials has the potential not only to reveal unique reactions to European imports but also to make more explicit the case-specific ideas, activities, and interests that helped set larger scale processes of material and social change in motion. Such understandings provide valuable new insights into the role(s) that technology and technological change actually played within broader processes of material and social transformation as European influence intensified. At the same time, identifying the manners in which particular materials and products are (re)manufactured and used and the meanings attached to them extends our cognizance of and appreciation for the breadth of human technological creativity and inventiveness as it is played out in multiple contexts of individual and group self-interest and expression.

Within this framework, the goal here is to explain how the Illinois responded technologically to European-introduced copper-base metal (copper and brass) kettles and metal sheet when these materials became available to them in the mid-seventeenth century. In this analysis, technology is viewed as an inclusive *system* that encompasses all stages and dimensions of activity and perception involved in making and using objects, from material acquisition through artifact fabrication, distribution, use, and discard (after Kingery 1993). Analysis is focused on explaining why one particular material, copper and brass, may have been attractive to the Illinois, and how and in what contexts they obtained and worked with it to fabricate and use decorative objects in particular social and symbolic contexts. The technical, behavioral, social, ideological, and historical dimensions of these technological processes are integrated to reconstruct what is referred to here as the "technological style" of Illinois copper-base metalworking and use (after Lechtman 1977). By using this approach, the early contexts of material continuity and change in Illinois technological systems are brought to light.

In concentrating here on bringing to the fore the technological "style" of one particular culture group, I am less concerned about how aspects of that particular style (operational sequences, technical choices) might differ from those inside or outside native interethnic boundaries (after Lemonnier 1986, 1993; Stark 1998). Rather, I am interested here in identifying and understanding how these and other features (as elements of larger technological *systems*) are expressed in contexts of action within and among cultures that are becoming increasingly exposed to European influences. It is these kinds of native responses that continually condition, redefine, and restructure native technological systems and social practices and thus, the ways in which material and cultural transformational processes take shape during this tenuous time.

ANALYTICAL APPROACH

The technological "style" of Illinois copper-base metalworking and use is brought to light using a sample of over 800 copper-base metal artifacts excavated from the mid-seventeenth-century Iliniwek Village historic site, Clark County, Missouri. The 40.5 hectare village site has been conclusively associated with the mid-seventeenth-century Illinois, one of the largest and most sociopolitically active native groups in the early historic period Western Great Lakes/Upper Mississippi Valley during the seventeenth century. They are one of the few peoples occupying the region during this period whose ethnic identity has been unambiguously linked with a distinctive and time sensitive material culture assemblage (Ehrhardt 2004; Ehrhardt and Conrad 1994; Grantham 1993).

Consisting primarily of sheet metal reworked into personal adornment items and the scrap from their manufacture, the copper-base metals industry from the Iliniwek Village represents one early manifestation of the initial phases of important processes of alteration, innovation, and change in Illinois technological systems. Converging lines of laboratory (archaeometric), documentary, and archaeological evidence are used to analyze and interpret the artifacts, reconstruct the technological system, and place the materials in their social and historical context(s). A suite of complementary archaeometric (laboratory) methods is utilized to reveal the manufacturing history and composition of the metals. These include metric/microscopic analysis, metallography, compositional analysis by proton-induced x-ray emission spectrometry (PIXE), and instrumental neutron activation analysis (INAA). Results from these investigations are integrated with contextual evidence from archaeological investigations and examination of primary and secondary historical and archival sources to document not only the range of technical processes involved in Illinois copper-base metalworking and use, but also to place the technology within the behavioral, social, and ideological contexts into which these appropriated and reformed materials were absorbed. In this way, significant new insights are gained on larger-scale questions of continuity and change in indigenous copper-base metalworking technologies and in the complex relations among material, technological, and social aspects of native cultural systems as new objects and ideas are brought into them during the earliest years of European influence.

At its core, this study is a structural analysis. While viewing particular material objects as important components of human material and social structural systems in a synchronic sense, it is also a study of structural dynamics as they pertain to technological systems. As such, it has both contextual and historical (diachronic) dimensions. This research operationalizes ideas not only about how the interrelationships between technological and social systems are played out in practice but also about how and why they change in response to particular historical conditions over the *longue durée* of history. It extends the boundaries of traditional structuralism in that it considers "structure" to include the design, meanings, contexts, and relevance of objects in particular everyday life settings of social action (Conkey 1989:52). In these frameworks, actions come out of the ideational realm and are products of ideological, symbolic, and value systems; these systems orient activity and are played out in practice (Cobb 1991:202; Hodder 1986:85; Sahlins 1981:7). According to Margaret Conkey (1989:152), these new perspectives on structural analysis "can elucidate how structures 'make sense' in particular historical contexts of social action."

However, this study takes structural analysis as "contextual meaning" (Conkey 1989; Hodder 1986) one critical step further. While it places these systems within their social settings, it also addresses what happens to them when large-scale potentially culture-altering historical conditions, events, and actions come into play within the long term. This is the diachronic dimension of structural analysis. Among others (Hodder 1986; Sahlins 1981; Trigger 1989 for example), Charles Cobb (1991) has posited that while systems are indeed structured over the long term, people are continually renegotiating and repositioning themselves with the passage of events (Dietler and Herbich 1998:246–247; Hegmon 1998:269). As members of open systems, individuals and/or the groups to which they belong are devising specific and particular forms of negotiation and relations within "conjunctures" or "smaller cycles of social histories that frequently alter structure to reach new points of equilibrium within the longer-lasting structural cycle" (Cobb 1991:171 after Braudel 1972:20–21). Cobb (1991:173) goes on to say that these continual transformations take place within groups and also in their external relationships "at various spatial scales." Importantly, he makes the point that "preexisting conditions of technology and demography also represent important constraints upon the course of human action" (Cobb 1991:205).

In this way, technology is not seen as a *constraint* on social action and process. Rather, it is viewed as a dynamic, active force that contributed in no small way to redefining and reorienting social practice in ways that shook native structures to their roots. How, why, and in what contexts this happened initially are the major avenues of inquiry here. This study, then, is a focused investigation of one particular element of the structural system, the technological system, as one nexus of intense activity at the "conjuncture" of European contact. In a general sense, it is an analysis of the domains, meanings, and contexts of technological activity over the *longue durée,* but it focuses specifically on the "conjunctures" and "events" of shorter-term history. The conjunctures of which I speak are, of course, related to European contact. These new influences and relationships generated multiple conditions for change that occurred in varied form and in multiple domains. Technology and material exchange is one of the major spheres in which these transformations took place (see Dobres and Hoffman 1994:215; White 1991). As an escalating, multilayered historical process, we see native actors negotiating and renegotiating to find new "equilibrium(s)" as altered structures and relationships emerge and are transformed again and again at a seemingly dizzying pace.

Native responses to these conditions should not be assumed to be uniform or universal, but should be recognized for their multiplicity. They need to be investigated in their specific historical and ideological contexts (Rogers 1990). This technological study of the record of early Illinois material response is one example of the many ways in which structures of materials selection and use are altered and social reproduction and renegotiation occurs in a contact context. It is particularist in the sense that it is culture specific, but at the same time, it addresses larger-scale questions about how, why, and under what conditions (contexts, motives, and intentions) technologies change as social production(s) is(are) continually being reconfigured and renegotiated. The processes identified here can then be examined in light of findings concerning alternative "reformulations of material practice" seen elsewhere in the native Northeast and in the New World (Sahlins 1981:7).

HOW THIS RESEARCH UNFOLDS

This investigation has four main parts. Chapters 2 and 3 develop and articulate the perspectives on technology as a "system" and technology as "stylistic." These ideas form the backbone of the study. Chapter 3 lays out the analytical program that operationalizes these perspectives. Chapters 4 and 5 provide contextual background critical to the research; chapter 4 presents a synthesis of Native American use of copper-base metal from prehistory through early historic times. It situates our understanding of protohistoric appropriation of European-derived copper and brass materials within that extended trajectory of technological and social history. Chapter 5 introduces the seventeenth-century Illinois in geographic, historic, and archaeological context. It relates the Iliniwek Village site's connections to the protohistoric Illinois and reviews briefly the results of excavations there.

The materials analyses follow in chapters 6 and 7, which present the results of the archaeometric and archaeometallurgical investigations. These findings define the range of copper-base materials at the site, characterize reliably their compositions and sources, and bring to light the range of production processes involved in transforming them.

In chapter 8, all of these lines of material, archaeological, and documentary evidence are drawn together to reconstruct and make explicit the ways in which the protohistoric/early historic Illinois responded technologically to the availability of European-introduced copper-base metal. In elucidating their particular "style" of response, I identify loci of adjustment, alteration, innovation, and even opportunity. Within this context, the role(s) that materials and technologies played in early contact period cultural dynamics are illuminated and made meaningful. An enriched picture of social action emerges as the Illinois resituate themselves and renegotiate their positions in an increasingly complex and rapidly changing world.

2 / Setting Aside the "Standard View"

Revealing "Style" and Change in Technological Systems

Native attraction to and consumption of a growing array of foreign merchandise set in motion important processes of material, economic, technological, and social change in indigenous communities. How archaeologists have explained those dynamics has changed significantly in the last decades. This chapter chronicles those changes with a view toward assessing how the role of technology(ies) has been portrayed in them. Materials-oriented models privileging (either explicitly or implicitly) the superiority of European goods and technologies as pivotal, even driving forces drawing hapless, helpless native peoples into an inevitable and inescapable downward spiral of material, economic, and social dissolution have largely been put aside. Recent revisionist perspectives place native peoples, rather than technologies, at center stage as central figures consciously forging their own unique paths of entanglement with foreign peoples, goods, and materials. Motivations for obtaining, using, even resisting European products are increasingly recognized as culturally and historically specific. Outcomes are no longer presumed to be regular or predictable.

These new positionings invite researchers to revisit and resituate the relations among native industries (native technologies) and the availability of European objects and materials on the one hand, and material, technological, and cultural change on the other. Our goal has been to develop new ways to examine and bring to light the variable, rather than determinative, relations among people, material objects, technologies, and technological change in settings of culture and material contact. This chapter sets out a comprehensive approach to identifying and interpreting the contexts of early material culture change into which analysis of materials can be woven. It forms the framework for the remainder of this study.

Drawn from ideas from the history of technology, the social anthropology of technology, and material culture studies, the approach taken here views technology as a dynamic and open "system" that considers all aspects of the technological process, from materials acquisition, to production, use, reuse, even discard (after Kingery 1993). Two aspects of this approach are notable: 1) while object-centered, it is not production-centered; production is merely one or more steps in a larger overall process of designing, making, reworking, and using things, and 2) all components of the system are evaluated within their behavioral, social, symbolic, perceptual, and historical associations and contexts (Figure 2.1). Taken together, the dynamics of particular technological systems as they are elucidated for individual culture groups is referred to here as "technological style" (after Lechtman 1977).

THE POSTCONTACT FATE OF NATIVE TECHNOLOGIES—THE STANDARD VIEW

Bryan Pfaffenberger (1992) begins his important review entitled the "Social Anthropology of Technology" by making explicit several widespread, but long unquestioned, assumptions concerning technology's place in human culture and how technological change takes place within it. He calls this the "Standard View." He is careful to point out that the Standard View is a "mythical" construct, and that not all researchers have espoused or agreed with all elements of it. Despite these caveats, he maintains that the following concepts have, until recently, dominated popular and scholarly thinking on the relations among material culture, technology, and culture: 1) necessity is the mother of invention, 2) form follows function/style and meaning(s) are secondary, 3) *man*'s (his purposeful choice of the androcentric term) material record chronicles a unilinear progression of technology from simple to complex, and 4) in the modern world, industrial technology (machines) controls society (Pfaffenberger 1992:494).

Pfaffenberger (1992:510–511) then expands this Standard View to consider what researchers have often assumed happens when industrial artifacts or modern technologies are introduced into "traditional" societies. According to this "view," after they receive modern goods, traditional cultures disintegrate and are forever changed by the impact of new technologies (see also N. Thomas 1991:2). By accepting "modern" technology, their systems are thrown into imbalance and are shaken to their very foundations. They become caught in the grip of technology and consumerism, which then shapes, even determines their futures.

The ideas Pfaffenberger brings out in this Standard View are reflected in many of the frameworks researchers in native northeastern North America have traditionally used to interpret material and social change in native cultures after European products are introduced. These points and perspectives might be termed the "standard view" of postcontact culture change in native northeastern North America. According to this view, when introduced to European goods through trade and gift giving, native response is immediate. Indigenous peoples recognize European materials and goods as technologically superior to their own, and instantly see the irresistible advantages in adopting them. This results in pressing demand for more European products, which quickly brings about hopeless dependency upon them. Soon, European goods replace native crafts. Traditional industries are abandoned and skills associated with them are quickly forgotten. Native need for constant inflow of new kinds of European commodities brings about major, irrevocable changes in traditional economic patterns and social relations as native peoples devote more and more energy to providing resources for the trade in order to procure these products. As European goods are adopted, so are European lifeways. Indigenous peoples are converted to Christianity. Ultimately, native cultural identity is completely forsaken as European customs, religion, and technologies inevitably pervade every aspect of native life. It soon becomes completely transformed (see Bradley 1987:2; Branstner 1992:178, 187–188; Nassaney and Johnson 2000:13–15; Rogers 1990:9–12; Rubertone 1989:344–45; Wagner 1998:432–434 for critical discussions of this perspective).

Although the "standard view" presented above is "mythic" as well, a number of ideas contained within it appear in a particularly powerful kind of acculturation model in which native peoples and their cultural systems have been viewed as nonresistant, increasingly dependent, and ultimately doomed as a result of sustained exposure to European ideas and products (see Morantz 1992 for a critique). Adoption of and dependence upon "superior" European goods are seen as a prime mover rather than as a variable in this scenario of cultural response. Cultural breakdown and identity loss rather than continuity and/or persistence are emphasized. The outcome of the process is portrayed as self-evident, predetermined, and inevitable.

PROBLEMS WITH THE STANDARD VIEW ACCULTURATION MODEL

Through the mid-twentieth century, *acculturation* became the interpretive paradigm for analyzing and explaining the processes and trajectory of post-contact culture change in native North America (Cusick 1998:126; Wilson and Rogers 1993). Its use was widespread both in anthropology and in archaeology (Cusick 1998; Fowler 1987). First presented in 1936 as a "Memorandum for the Study of Acculturation," authors Robert Redfield, Ralph Linton, and Melville Herskovits (1936:149–152) defined the concept as "comprehend[ing] those phenomena which result when groups of individuals having different cultures come into continuous first-hand contact, with subsequent changes in the original cultural patterns of either or both groups." They go on to present ways to approach and analyze the problem, focusing on culture "traits" and potential conditions, mechanisms, processes, and results of their transfer from one group to another. Over subsequent years, the ways in which anthropological and archaeological models of acculturative change have been applied are extremely diverse. However, certain assumptions about the meaning, direction, and outcome of culture contact regardless of context have come to be associated with some formulations of the model. As a result, acculturation as a whole has come under severe criticism by both anthropologists and archaeologists.

Loretta Fowler's (1987:6–9) critique of acculturation approaches as they have been applied to native, particularly to Plains, groups in anthropology has immediate relevance for a discussion of the "standard view" of post-contact culture change presented above. First, anthropologists once tended to see acculturation, or the extent to which native people have given up their own way of life and adopted a European one, as a matter of degree—cultures are viewed as "more" or "less" acculturated (Fowler 1987:6). Degree of acculturation was measured by the extent of foreign "trait" acceptance—the more white-oriented traits a group exhibits, the more acculturated they are. Fowler points to the inappropriateness of such an approach, as traits are often taken out of their cultural context. Out of context, traits are often not understood by anthropologists the way native actors have interpreted them within their own systems of meaning. Anthropologists may "read" particular behaviors as indicators of acculturation, whereas the actors themselves see newly integrated "traits" or "ideas" as expressive of new aspects of their *own* rather than of a new or different cultural/ethnic identity. Appropriation of particular "traits" may be responses to new challenges and opportunities brought about by new cultural or lifestyle situations. They might be considered an aspect of culture continuity rather than acculturation within groups whose cultural identity(ies) and traditions are constantly being reconceptualized and reinterpreted (Fowler 1987:7–8).

Second, in acculturation models, Native Americans are often seen as passive acceptors of new lifeways rather than as creative and resourceful decision-makers in bringing about change and ultimately shaping their own futures. Native life before contact is seen as static—living but stagnant, unchanging traditions virtually without a historical dimension to condition motivations for and reactions to change (Fowler 1987:7; see also Ramenofsky 1991:439–440). After contact, traditional culture loss as a result of sustained influence is assumed, and assimilation to European ways is the rational and expected, rather than possible, outcome.

Third, native/white relations are privileged over native/native ones, thus obscuring the role intergroup relations and the exchange of ideas and strategies may play in bringing about or mitigating change (Fowler 1987:9). Also to be considered here are relations within the group and the varied interests of individual members potentially acting as agents of change.

Critiques of acculturation models in archaeology largely mirror Fowler's for anthropology. J. Daniel Rogers (1993) offers two important points. First, the notion of acculturation as it has commonly been adopted is ethnocentric in that it "assumes a directed outcome" (although he is correct to point out that predetermination of outcome was not explicit in all formulations). Second, in an archaeological correlate to Fowler's critique, cultures are "typologized" (after Quimby and Spoehr 1951, and J. White 1975), that is, they are placed in categories according to the degree of acculturation they exhibit, thereby obscuring the possibility of understanding the complexities and repercussions involved in various kinds of interaction situations (Rogers 1993:74–75).

James Cusick (1998:135–138) makes equally relevant arguments in his detailed critique of the history of acculturation and the use of such models in anthropology and archaeology. Like Fowler, he observes that in these models, Western people are considered active agents, while indigenous (non-Western, contacted peoples) populations are considered passive recipients. Again reiterating Fowler, he emphasizes that archaeologists have confused changes in material culture and in behavior with culture change and changes in cultural identity. Here, he echoes other authors who have expressed concern with uncritically assuming that changes in material culture signal changes in the nonmaterial realm (Rogers 1990:226). Also for Cusick, acculturation frameworks are faulty in that outcomes are predictable; in actuality, particular aspects of culture may change in different ways in the same or different circumstances. Trait list approaches and trait-driven interpretations, no matter how well contextualized, are an ineffective means of analyzing culture change (but see Barber and Berdan 1998:190–191 for a form of their continued use by ethnohistorians). Lastly, Cusick emphasizes that acculturation models do not take power relations, especially vastly unequal ones, into consideration. Only recently have notions of native resistance and creativity in response to particular circumstances of encounters been reconsidered (see Rubertone 1989; Saunders 1998).

Although critical of many aspects of acculturation approaches, Cusick stipulates that with critical application and a thorough understanding of its shortcomings, many formulations of acculturation may still be useful for archaeologists (Cusick 1998:126). Edward Schortman and Patricia Urban (1998: 107) claim that interaction theories must "draw on but transcend" many aspects of both acculturation and world systems frameworks in order to adequately represent the dynamics of culture contact.

MATERIAL CULTURE, TECHNOLOGY, CULTURE CONTACT, AND THE STANDARD VIEW

In the Northeastern Woodlands and Western Great Lakes, the relationships among indigenous technologies, technological change, and social change in contact contexts have been modeled within the "standard view" framework in different ways. While many formulations appear to focus on European *products* and their role in native cultural transformation and disintegration, it is important to emphasize the often implicit, but powerful *technological* axis on which many of them actually turn. When examined from such a viewpoint, technology as "the means by which man [*sic*] controls or modifies his natural environment" is actually the force that underpins material cultural change (Spier 1968:131). For fur trade specialists like Harold Innis, European products, especially iron tools and kettles, are thought to be desired because they are superior to native-made counterparts (Innis 1970 cited in B. White 1994:370). Technological "superiority" on the part of the donor culture and "improvement" on the part of the recipient culture are assumed. Framed in this manner, technology is viewed as a determinant rather than a variable in scenarios of initial material culture acquisition and eventual postcontact dependency.

Indeed, the initial Native American/European contact experience has been referred to as a "clash of technologies" (Rogers 1990:10) or as a "technological invasion" (Martin 1975:112) in which, according to Bruce Trigger (1986:255), significant asymmetries in technological systems are emphasized. Why is technology figured as central in these models? Linda Cordell and Vincent Yannie (1991:96) posit that in general, archaeologists' long-term interest in technological (and environmental) aspects of the material make-up of the archaeological record stems from greater familiarity and experience with these components, as well as their amenability to scientific methods of analysis. Timothy Earle (1991:83) claims that from a New Archaeology perspective, factors from outside the cultural system such as technological change are frequently cited as "prime movers" in the search for explanations for cultural change. Like Earle, Charles Cleland (1992:77) suggests that perhaps archaeologists working with the material record tend to overemphasize the role technological change plays in sociopolitical transformation. Such perspectives may have been reinforced by the ideas of neoevolutionist Leslie White concerning the principal role of technology and technological change in shaping social organization and cultural ideologies (Spier 1968:132; Trigger 1989:22). Taking a reflexive stance, Barbara Bender (1989:84) has proffered that the notion of technology as prime mover (or precondition) in processes of social change stems from our experience in an industrialized West, where technology is seen as a primary catalyst for change and where increased social complexity and technological change are found to go hand in hand (see also Pfaffenberger 1992:510).

These observations are directly applicable to this investigation of assumptions concerning Europeans' presumed technological superiority and resultant native dependency. Within an adaptationist framework, James Fitting (1976) ties the adoption and subsequent dependence on superior French goods to an increased level of native "fitness" or adaptation. Using L. White's formula relating group technology and energy use to level of cultural development, Fitting (1976:329) determined that one European iron tool equals in efficiency 23 native-made stone counterparts, and that one brass kettle equals 60 ceramic vessels. Claiming that these superior items increased Native American technological efficiency within their already viable cultural systems, Fitting (1976:333) concludes that, "The French helped the Indians be better Indians."

The central place that technology has been afforded in Fitting's model may well have roots in adaptationist explanatory frameworks. In his model, selection of technologically superior items is seen as a function of indigenous peoples' needs—or wants—which, in turn, increase their adaptive fitness to the environment. Pfaffenberger (1992:496) correctly aligns these explanations of need-driven technological change with the ideas of Lewis Binford and the New Archaeology (Binford 1962, 1965). From this perspective, technoenvironmental factors and technological change (adaptation) are viewed as primary in maintaining group survival and initiating social change (Trigger 1989:24). Actually, however, the "needs" of which Fitting speaks might be better framed as "perceived" needs rather than biological or "natural" ones. According to George Basalla (1988:6, 13–14), humans "cultivate" more and more complex and unnecessary ways to meet our "perceived" needs, which have nothing really to do with our ability as humans to survive.

It might also be argued that Fitting's "better tools make better Indians" argument runs *counter* to the Standard View's directional acculturation or "Europeanization" outcome as Indians still maintain their cultural identities, but with European tools they are "better" at it. Patricia Rubertone (1989:36), however, claims that what Fitting and others are really saying is that "a more European-like Indian was . . . a better Indian." She claims that Native Americans are inappropriately depicted as "copiers" or "imitators" of European technologies. As uncreative absorbers of European goods, they did not play active parts in producing their own cultural histories (see also Trigger 1980). Such a perspective precludes thinking about native initiatives and intentions with regard to material culture and social concerns (Rubertone 1989:37–46).

ALTERNATIVE PERSPECTIVES TO THE STANDARD VIEW

Finding the "standard view" (embedded as it is in acculturation theory and turning on technological determinism as well as perceived Native American need fulfillment) unsatisfactory, how have archaeologists begun to rethink the problem of deciphering native response(s) to European materials? How can the relationships among native and European materials, technologies, and social change be repositioned within that problem? Northeastern North American archaeologists have only recently begun to rethink issues of sociocultural change in a culture contact context from a material culture or technological perspective (Bradley 1987; Branstner 1991, 1992; Hoffman and

Dobres 1999:209). Many have looked critically at some of the assumptions that were either implicit or explicit in the "standard view" model. They have begun to reexamine the impact of contact on native peoples not in terms of "acculturation," wherein material culture change is considered to be a direct and simplistic "record of culture change," but as a complex suite of historical and cultural processes in which Native Americans were thoughtful decision-makers in the selection, modification, integration, or rejection of new objects, materials, and technologies into their systems of value and activity (I. Brown 1979:112; Quimby 1966:8–9; Rogers 1990:11, 13).

Ethnohistorian Richard White (1991) makes a cogent case for this position as it pertains to native peoples of the *pays d'en haut* (or upper country, which roughly comprised the lands surrounding the Great Lakes west of Iroquoia to the Mississippi) in the seventeenth and early eighteenth centuries. R. White (1991:99) argues that native "desire for goods was not automatic or simple." He (1991:95–101) claims that native peoples engaged in trade for exotic materials not for commercial gain or for prestige/wealth (a juxtaposed European view) but because foreign goods had special value for them in exchange as gifts in diplomatic and social situations with both Europeans and with each other (see also Helms 1993). According to R. White (1991:101), their bestowal "created reciprocal obligation and established status." Goods were procured by native peoples and circulated in a variety of settings primarily to meet *social* needs. They thereby became *cultural* rather than *material* or *survival* necessities. Forging and reinforcing relationships and alliances by means of material exchange was an important element in maintaining the structure of sociopolitical relations and also served as an important reason to uphold the structure of the trading networks. In the *pays d'en haut,* then, social interaction and material trade and exchange went hand in hand.

At the same time, R. White (1991:128–136) maintains that in the interior, long-established subsistence patterns and material/technological repertoires did not change quickly as a result of the initial appearance of trade goods. He suggests that while trade goods had both utilitarian and symbolic value, they were rare until well into the eighteenth century. As such, they were in too short supply to supplant native-manufactured items entirely, obviating the suggestion of native "dependency" (see also Morantz 1992). Should avenues for acquisition of merchandise have been cut off, native peoples were still well able to fall back on their own subsistence practices.

Michael Dietler (1998:300) has also argued that in many colonial exchange situations, "European goods were no means inherently irresistible to indigenous societies. Usually these [indigenous] peoples exhibited highly selective preferences in goods they were willing to accept and to give in exchanges with colonial agents, and they sometimes refused to interact at all." He (1998: 300) provides the following example from the seventeenth-century native Northeast:

> [E]arly English attempts to engage the Native American tribes of New England in the fur trade were a failure because those peoples were not interested in the various goods the colonists had to offer; it was only when the English finally discovered the native demand for *wampum* that they were able to begin an exchange relationship.

How foreign materials and goods are used in native contexts and become integrated into indigenous social systems is now seen as historically, culturally, even cosmologically conditioned. As Mary Helms (1993:105) has so cogently argued, drawing new things acquired from a distance (outside) into a new "home locale" (an ordered social heartland inside) and transforming them is an extremely complicated material and nonmaterial process. J. Daniel Rogers (1990:10) argues that neither superiority in utility (see Fitting 1976:333) nor economic value alone can explain why one product or technology might be selected over another. Decisions to accept or reject items also involve social and ideological factors such as symbolic meaning and world view, as well as the material and technical context of intended use.[1] As Nicholas Thomas (1991:87) has noted, in situations of material appropriation "technology is dependent upon cultural knowledge: even relatively specialized tools do not have specific purposes inscribed in them, and purposes and uses are variously relevant and recognized."

As many researchers both within and outside of anthropology have noted, technological activity is a complex social process (Dobres 2001:48; Dobres and Hoffman 1994:247; 1999:211; Holl 2000:21–24; Hughes 1990; Pfaffenberger 1992, 2001:77). With regard to transformations, Merritt Roe Smith (1985:164) claims that "social forces not only enframe but actually define technological change." Thus, the successful transfer of new suites of techniques, knowledge, artifacts, and materials, which often brought about the abandonment of traditional counterparts, was the result of purposeful decisions into which all of these historical, cultural, and technical considerations were likely woven (Branstner 1991, 1992; Hamell 1983, 1986; Miller and Hamell 1986; B. White 1994; R. White 1991:13–15). From this perspective then, change in any aspect of native material systems should not be considered trivial or superficial (but see Bradley 1987:174–175).

Thus, native groups are no longer seen as either "becoming better Indians" through "adaptive," need-driven appropriation of European items, or transforming "into" Europeans through their ever-increasing consumption of them. Rather, they are now recognized as creative and innovative in their approaches to European products and materials (Rubertone 1989:36). As Ian Brown (1979:113) and others (Bradley 1987; Pfaffenberger 1992:511) have noted, European objects were often selected in one form, transformed and reinterpreted in new or traditional ways rather than in European ways, then used in contexts very different from the ones originally intended. As William Schaniel (1988:493) points out, "the adopting of introduced materials does not imply the concurrent adopting of the system of logic that produced the technology." It can no longer be assumed that by appropriating European technologies and material culture, native people lost their cultural identities and group definition (Fowler 1987; see Wagner 1998:430–431 for a Western Great Lakes example).

This is not to deny that native cultures were deeply affected by the material and technological changes that took place even in the earliest years of engagement with Europeans and their products; like many before him, Cleland (1992:77) postulates that it happened rapidly in relation to changes in other spheres of native life (Spicer 1961:542; J. White 1975:160; Wolf 1982:161). Patrick Malone (1991:31–32) claims that with natives' appropriation of firearms within their own style of warfare, its organization changed significantly. In turn, English adoption of native peoples' "skulking" way of forest warfare changed foreign tactics as well. Aubrey Cannon chronicles how metal beads and ornamentation became an important form of competitive display involving both Huron men and women (Cannon 1991). Calvin Martin (1975:113–114) and a host of others have documented the abandonment of aboriginal ceramic and lithic technological skills and knowledge as native peoples took up use of European "functionally equivalent" kettles and knives (Brose 1971:55; Walthall 1992:156).

Important also are the ways in which native peoples reinterpreted new foreign materials and implements. As Rogers has said (1990:108), although native consumers were desirous of some elements of European culture, they consciously eschewed many other aspects or objects that were not considered compatible within their internal cultural logic. According to ethnohistorian Bruce White (1994:369, 376), at first, Western Great Lakes native groups admired and appreciated Europeans for the power of their technology. They thought beads, guns, knives, and cloth to be magical. Guns, kettles, and glass beads were valued for their intrinsic spiritual "power" rather than their increased effectiveness in a practical realm. Only later as European goods became more abundant did Europeans and their products lose their exalted stature (Bradley 1987:169; B. White 1994:371). George Hamell (1983, 1986) and later Christopher Miller and Hamell (1986) have argued quite convincingly that native peoples began using glass beads and imported copper-base metals not because they were technologically more efficient than native counterparts, but because of their similarity to objects and materials already important in their own native symbolic worlds.

NEW PERSPECTIVES, REVISIONIST INQUIRIES

These shifts in perspective have not resulted in the emergence of one inclusive theory of culture contact to replace acculturation as the single overarching explanatory framework (Cusick 1998:127). As in Spanish Borderlands research (D. Thomas 1989:6; 1991:xvi–xvii), many new alternative materials-based ways of looking at the processes and consequences of material and cultural transformation in situations of culture contact have emerged in Northeastern Woodlands and Plains research. Noteworthy examples include James Bradley's (1987) chronicle of the tempo and trends of sixteenth- and seventeenth-century Onondaga material appropriation and concomitant cultural adjustment; J. Daniel Rogers' (1990) study of the processes and sequence of material culture change as conditioned by the evolving value and logic systems of both the Arikara of the Upper Missouri River area and the Europeans from the late 1500s through 1862; Patricia Rubertone's (1989) demonstration of the means through which wampum, as an aboriginally produced technological product, was consciously used by the Narragansetts to resist British economic and colonization goals; Susan Branstner's (1991, 1992) use of decision-making theory to challenge traditional interpretations of early acculturation among the Tionontate Huron and to reveal arenas of material culture change and persistence; and Ann Ramenofsky's (1998) explanatory recast of the processes of native Overhill Cherokee and the Mandan-Hidatsa material change over the contact period in Darwinian evolutionary (selectionism) terms. For the Western Great Lakes, Mark Wagner (1998:432) has reviewed how both archaeologists and social historians have reframed postcontact material and cultural change within gender studies, Native American agency, Native American identity, cultural brokerism, social and kinship relations between Native Americans and Euro-Americans, and the consequences of increasing participation in the fur trade.

Set in a number of different historical and cultural contexts of European contact and representing a variety of perspectives, these studies share one common thread—detaching from traditional "standard view" acculturation frameworks. They reflect a realization on the part of an ever-increasing number of researchers that acculturation-type constructs for explaining culture change are, as Dietler (1998:288) has said of those applied in the ancient world, "inadequate because they are structurally overdetermined, teleologically reductionist, and incapable of accommodating local agency." In setting aside these frameworks, archaeologists are increasingly demonstrating that many factors condition the timing, tempo, and trajectory of material and culture change in particular cultural settings. These factors include the diverse types and social conditions of intensifying interaction, the differing geographic and historic backdrops of encounter, the various and changing motivations of both foreigners and Native Americans, and the often unpredictable responses of the participants. No one would now deny the pivotal role that the introduction of new material culture and new technologies played in instigating technological change or in setting off larger scale processes of cultural transformation, nor would they question the value of elucidating these changes at the level of individual cultures (but see Deagan 1998:27 for a discussion).

ASSESSING "TECHNOLOGICAL CHANGE" IN EARLY CONTACT MATERIAL CULTURE ASSEMBLAGES

Now that technologies (both European and indigenous) and technological change have been effectively repositioned as variables in the dynamics of early cross-cultural encounter, how might they be recognized in contact period material assemblages? What does the initial appearance of European objects and materials in these assemblages actually *mean* in terms of evaluating the nature, extent, and contexts of technological impact in contact contexts? The extent to which early adaptation and use of European items appearing in material culture assemblages actually represents *change* in native technological repertoires and domains of social action becomes a crucial question. Trigger

(1987:243–245) has claimed that the appearance of trade goods on Huron sites through early, indirect trade in the late sixteenth and early seventeenth centuries had *negligible* effect on native technologies. While he acknowledges the appearance on native sites of iron and copper-base metal objects and beads manufactured from sheet and broken kettles, and an increase in number and kind of goods over time, he also posits (1987:243) that "the most obvious change . . . is the inclusion of trade goods and other items in ossuaries for the first time." Further, Trigger (1987:244–245) asserts that from a native point of view, early trade in exotic goods and its transformational impact on native political and external exchange systems was much more significant than the changes brought about by how new objects functioned within the recipient societies themselves.

Here, Trigger looks outward, privileging the expansion of external trade relations while paying much less attention to *why in the first place* people may have been drawn to the very objects that precipitated the trade expansion of which he speaks. We might logically ask how and why trade would or should expand if potential consumers did not have particular, culture-specific reasons (not the least of which might be technological ones) for wanting to take in what was available to them? By concentrating on the large-scale changes in regional and interregional political and exchange structures, Trigger short-changes the everyday, local contexts and processes of early material consumption, appropriation, and use among specific cultural consumers. These everyday processes not only drive and sustain the macroscale changes to which he refers but also serve to bring about alterations in material culture and social systems at the level of individual cultures.

Two additional studies focusing on assessing early material culture change in individual cultures are highlighted here. In his thoroughgoing study of Onondaga Iroquoian adjustments to European influence over time, Bradley (1987:5) claims that material culture and material culture change are "particularly sensitive" means through which evolutionary change in postcontact native culture can be measured. However, his assessments of what actually constitutes *change* in native material production and use are of interest here. He points to two instances at Onondaga in which, by the mid-seventeenth century, European objects had replaced native functional counterparts; iron axes had taken the places of both lithic and iron celts/adzes, and copper kettles had replaced native-made pottery. Having made this important observation, Bradley (1987:175) concludes that since "change" seems to amount to the adoption of functional substitutes, he terms it as

> superficial . . . changes were more a matter of grafting new materials, forms, and perhaps some functions onto an existing set of preferences, not replacing one set of values with another. In fact, at the level of basic cultural beliefs and values, it is questionable how much change had really occurred by the mid-seventeenth century.

At the same time, however, he identifies other early contexts of intense material activity and elaboration that hardly seem superficial. New native craftwork appears, yet traditional native crafts and material use do not disappear. Copper-base metalworking emerges with the availability of European metals. The manufacture of metal implements and ornamental items becomes more refined through time, and even extends to the adoption of casting techniques. Despite these changes, work in lithics or clay does not fall away altogether; stone tool production shifts to making gunflints, and work in clay switches from pottery manufacture to clay smoking pipe production. Importantly, Bradley further notes that much native work in many traditional materials was actually amplified and became more innovative. Yet, for him, (1987:178), because these changes are focused largely on change "internal to Onondaga culture and with limited applicability beyond it," he considers them trivial. Only at a later time, when interaction intensifies and there is more evidence of "mixing and blending . . . of European materials, techniques, and concepts" does cross-cultural material influence become significant in Onondaga culture (Bradley 1987:178).

In her study of material and technological change among the Tionontate Huron, Susan Branstner (1992) has argued that in many instances, European materials were substituted for native items or manipulated and reshaped into traditional forms and then used in traditional ways. For instance, copper-base metal wire was shaped into ornaments, and kettles replaced native-produced clay pots for use as cooking vessels. In the case of native selection of kettles as clay pot replacements, she reiterates Fitting's assessment that loss of traditional ceramic production skills and dependency on European items quickly ensued once kettles were introduced (Fitting 1980 cited in Branstner 1992:191; see also Trigger 1991). Glossing over these processes, she argues that adoption of such functional/technological equivalents does not alter the "meanings" and "goals" of the native technological or social system; therefore they cause little "dramatic" change in the system itself (Branstner 1991:48, 248; 1992:183, 187, 191). At the same time, however, she (1992:191) asserts that "the selection of metal containers over ceramic suggests there may have been a reallocation of the pot maker's time for other duties. Perhaps the Huron women were spending more time producing a corn surplus for trade." For Branstner, alterations in women's work allocations to "spend more time producing a corn surplus for trade" appear not to constitute change in technological activity or an alteration in technological goals. For Branstner, decisions to replace objects such as clay pots and bone needles with items that perform what she calls the "same technological function" are based on "efficiency, durability, and availability." It involves purposeful choices made "as solutions to problems" (of efficiency, durability, and availability?), but since these decisions do not alter "goals specific to each context" they *do not constitute or necessitate technological change* (1992:187, 191).

The extent to which the above researchers see or do not see the important ways in which even the most "trivial" responses to new materials acquisition and use can set in motion early processes of technological and social transformation may reflect a problem with how they conceptualize the term "technology." Trigger (1987) and Branstner (1992) both view technology in terms of the *functions* European or native-transformed objects assumed as they were assimilated into native life. When he is explicit, Bradley (1987) refers to it only in terms of artifact *production.* None of these perspectives is adequate to explain the potential range of social and technical behaviors, activities, and attitudes involved in all stages of the technological process, from new ways to procure raw materials, to the new skills and knowledge needed to manipulate or "recycle" materials, through the diverse situations and consequences of use and reuse, and finally to contexts and possible meanings of or reasons for discard. Consequently, contexts and activities constituting change are overlooked.

CONCEPTUALIZING TECHNOLOGIES AS DYNAMIC "SYSTEMS"

There are myriad approaches that could be taken in the search for a view of technology suitable for interpreting the complex relationships among native industries and the dynamics of change and continuity in those industries in the contexts of early material transfer. However, as we have already determined, a broad-based one that builds on the notion that technology and technologies are embedded within culture and are socially and culturally dictated is most productive (see Dobres 2001:48–49; Dobres and Hoffman 1999;

Schiffer 2001). Thomas Hughes's conceptualization of technology as a "seamless web" sociotechnical system in which social forces shape technology and technological change has great appeal since it considers the pervasive, inclusive nature of technological activities and processes as cultural productions (Hughes 1987:13, 17; 1990:7, 24; Pfaffenberger 1992:498).

Marcia-Anne Dobres and Christopher Hoffman (1994:199) expand on this idea well for use here. While aware that technologies involve action on materials (after Lemonnier 1986, 1993), they also embed these activities in *historically specific* social action, worldview, human agency, and social reproduction. For them, contexts of activities, social, political, technical, and symbolic aspects of materiality, and the relationships of technological knowledge to social knowledge and action are all important aspects of understanding technology. Human agency is at the fore in all of these contexts (Dobres and Hoffman 1994:215; Hoffman and Dobres 1999:2–4). As they (1999:3) have recently argued,

> to conceptually limit our definition of technology to tangible, utilitarian, and practical object matter because it is seemingly more knowable is to cut off from that consideration an understanding of the dynamic subject matter that makes human technologies both possible and meaningful. That subject matter, of course, is people, their social relationships in productive endeavors, and their (not necessarily uniform) attitudes, values, and rules about making and using the material world.

Adopting these conceptual perspectives, if we then view technology or "a" technology as a "technological system," as W. David Kingery (1993:117–118) does, as "consist[ing] of materials acquisition and distribution, design, manufacturing, product distribution, reception, perception, and use as well as various possible reuse and discard technologies," we have located an inclusive, holistic means of approaching technological analysis that encompasses the entire range of technological activity and meaning surrounding making *and using* things (Figure 2.1). It does not limit analytical focus to design alone, production alone, or function alone; it recognizes that all components of a technological system are linked to and interrelated with all of the others. His conceptualization also incorporates all of a technology's dimensions, material, behavioral, social, ideological, and aesthetic, as integral and equally important elements of the system. He (1993:227) recognizes that innovation,

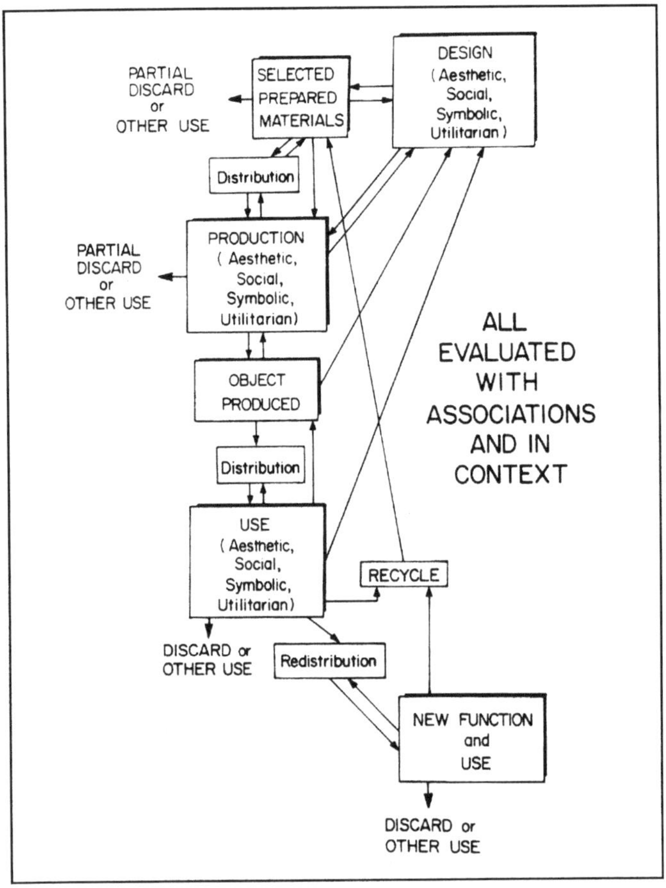

Figure 2.1 W. David Kingery's conceptualization of a technological "system" illustrating its inextricable roots in culture context and social action. From *History from Things,* edited by Steven Lubar and W. David Kingery, published by the Smithsonian Institution Press, Washington, D.C., copyright © 1993 by the Smithsonian Institution. Used by permission of the publisher.

novelty, even revolutionary technological change can arise from alterations in any of these dimensions or aspects of the process within culturally or historically specific conditions; changes in one aspect of the system influence the others.

Lastly, and importantly for my position here, Kingery (1993:218) cements the linkages between material culture, technological systems, human agency, and culture, arguing that,

> Each of the activities forming a part of this system—design, manufacturing, distribution, and use—requires human perceptions, human cognition, social organization, and human interactions as well as artifacts for its performance. This is more than context; it is an integral part of technology itself . . . through the use of artifacts, technology permeates all human activities and behavior.

Applying these new perspectives, a reexamination of Branstner's work, for example, would bring archaeologists to quite different conclusions about whether changes in material but not in artifact function actually constitute change within Huron technological systems. Alterations can indeed be seen at several points. The acts, attitudes, and choices involved in adapting and using European objects and/or materials, even if they are adapted to meet what are thought to be the same functional goals, *do* indeed constitute change within the "technological system" as Kingery has defined it. Changes occur in the dynamics of procuring the new materials/objects, in selecting one material/object over another, in acting on that material (in terms of manipulating the material and manufacturing finished objects), and in integrating the new components of the system through both social and symbolic acceptance and through use in the same or different contexts. Thus, even when "functional equivalents" or "substitutes" (glass or copper-base metal beads for shell beads, metal awls/knives for stone counterparts) appear in indigenous material repertoires, it cannot be assumed that they have negligible meaning or do not represent change in native technological systems.

Manipulating and reshaping European materials also signals innovation and change within important *technical* aspects of the system, that is, in terms of the gestures and operations, objects, and knowledge involved (after Lemonnier 1980 cited in Schlanger 1994:145). Ability to manipulate new objects and work new materials requires acquisition of new technical knowledge of their physical and mechanical properties, limitations, and capabilities in order to decide whether to accept or reject them and to determine the contexts of their potential use(s). In the case of manipulated or transformed materials and objects, it requires development of appropriate technical knowledge, skills, and the availability, appropriation, or manufacture of other task-specific tools (some of them new types perhaps, such as scissors or tin snips) necessary to perform actions or "operations" on materials in order to transform them successfully and achieve the desired functional and/or aesthetic result (Lemonnier 1993:6–7; Sigaut 1994:439–440). It is important to note, however, that solutions selected may not be the most practical or efficient ones available; social or ideological factors may constrain options, strategies, and/or choices. Choices and solutions may change over time (Lemonnier 1993:6–7, 16, 24).

Equally importantly, successfully integrating European materials and/or artifacts into native repertoires involves ideological, symbolic, and social acceptance of and/or adjustment to a new or exotic material or object based on native systems of value and meaning, which themselves may be in flux. Pierre Lemonnier (1993:18) cites several factors, including economic and political considerations, group or subgroup special interests, and status representation and differentiation, as important in rejecting or accepting new technologies or technological features. Aubrey Cannon (1991) provides an excellent example of this by highlighting the important role that the politics of gender relations played in Huron consumption of new materials. In addition, assimilation and new patterns of consumption of newly available materials may also cause new demand, bringing about changes in other aspects of the sociopolitical, subsistence, and settlement systems.

BRINGING TECHNOLOGICAL SYSTEMS AND TECHNOLOGICAL CHANGE TO LIGHT: A MATTER OF "STYLE"

I have proposed above that Kingery's technological "system" conceptualization of technology as an inclusive system fits well as a baseline framework for elucidating technological change in early culture contact contexts. It is especially relevant for use here because the "cultural biography" of materials and things becomes so complex in exchange, recycling, and reuse situations where objects may become commodified and "alienated" from their original producers and contexts of use, while at the same time maintaining an aura of the exotic (Helms 1993; Kopytoff 1986:67; Martin 1975; N. Thomas 1991:39; Turgeon 1997). In responding to and redefining these objects and materials, native peoples have integrated them into their own material, symbolic, and ideological realms. Kingery's conceptualization of technology as a "system" permits all of these dimensions to be highlighted as a part of what might be termed the "technological life" of things.

How then, might Kingery's conceptualization of technological systems become operationalized in material culture analysis in a way that takes in all of its coherence and its dimensions? Since 1977, materials scientist Heather Lechtman (1977, 1994) has presented a number of works articulating her concept of technological "style." Influenced by historical metallurgist Cyril Stanley Smith's work on structure and style, her approach is anthropologically and archaeologically focused. It centers on the notion that objects (as archaeological artifacts) are the physical manifestation of culture, and as such, make cultural statements that embody not only behaviors, but also belief systems (Lechtman 1977:4). These beliefs, or "symbolic relations," link artifacts and people. Lechtman acknowledges that artifacts are often studied in terms of their formal characteristics, the way they are used, what they mean, or how they are exchanged in and among prehistoric societies. However, she has concentrated on steering material culture analysis toward what might also be learned by investigating *how* objects are made and what the technologies themselves can reveal about deeper cultural patterns. One important way of doing this is through the use of laboratory techniques that bring to light their manufacturing histories. For Lechtman (1977:4), how objects are fashioned, that is, the behaviors, activities, processes, and patterns involved in making objects in specific cultural contexts, constitutes what she has called technological "style." She (1977:6) argues that these behavior patterns include multiple elements, including "technical modes of operation, attitudes towards materials, some specific organization of labor, ritual observances—elements which are unified nonrandomly in a complex of formal relationships." Both the technical actions themselves as well as the arrangement and sequences of operations communicate important underlying patterns of social belief and action that can be interpreted (Lechtman 1977:12). It is this "package" of technological activities as culturally patterned systems of ideas and behavior that is "stylistic" (Lechtman 1977:6; 1994:5).

Lechtman (1977:7) recognizes that technological style is "emic" in that it is culture-specific, reflecting technological decisions and action on materials potentially unique to time, place, and social context. At the same time, it is "etic" because raw materials behave mechanically and structurally in the physical world in specific, predictable ways that are discoverable. Determining the patterned behavioral and attitudinal relations between the two, and interpreting what is being expressed in the particular patterns we have identified is the point of elucidating technological "style." For Lechtman (1977:10–12), the goal in investigating "style" is to converge the ideational and the phenomenal by linking the patterns and rules of technological performance with the "attitudes of artisans towards the materials they used, attitudes of cultural communities towards the nature of the technological events themselves, and the objects resulting from them."

Importantly, Lechtman (1977; 1994:5) argues that technological style in prehistoric material culture manufacture can be brought to light and elucidated through a materials science and materials engineering approach to analysis. Artifacts, she maintains, contain precise information about the history of their manufacture. Materials have physicochemical properties that constrain and potentially alter them in manufacture and use. At the same time, they have other properties that are particularly compatible with and amenable to certain types of manipulations. Therefore, they present to the artisan particular problems that must be managed in order to produce a desired result. Often, problems of design, material, and fabrication have alternative solutions that involve choices on the part of the artisan. Using laboratory methods to reveal these properties and alterations brings to light their manufacturing histories, and thus reflects the particular ways artisans chose to solve these problems. Results also contribute significantly to reconstructing prehistoric technological sequences and processes of manufacture. Thus, her approach illuminates important aspects of technological style irrespective of social context of use (Lechtman 1994:5).

Results of Lechtman's own work (1977, 1994) on the development of New World metallurgy and metalworking style in the pre-Columbian Andes illustrate her perspectives. However, few other investigations of technological style in New World material culture or metallurgy followed (see Bradley and Childs 1991; Childs 1991a, b; Hosler 1994 for exceptions). Dorothy Hosler (1994) followed Lechtman's analytical stance and materials science approach, tracking development and expansion of copper-base metallurgy among several diverse New World (Mexican) prehistoric cultures. According to Hosler, the West Mexican metalworking tradition (A.D. 600–1300) was distinctly nonutilitarian, and emphasized the resonance and golden and silvery color of the metals as of primary importance in artifact manufacture and use. Metal was regarded as divine, indestructible, and sacred; it clearly communicated sacred power through the qualities of sound and color. These qualities could be created metallurgically. Golden and silvery metallic colors were associated with the property and divine emissions of solar and lunar deities (respectively) in Mesoamerica (Hosler 1994:228). Highly iridescent and reflective objects were metaphoric representations of the objects and beings inhabiting the Aztec paradise (Hosler 1994:232), where bell sounds were considered metaphors for human and bird song, birds themselves, flowers, trees, and golden colors.

A much smaller-scale but pertinent metals study by James Bradley and S. Terry Childs (1991) demonstrates the utility of such an approach in analysis of contact period metal artifacts. Commingling archaeological, materials science, and ethnohistoric research, Bradley and Childs reinterpret the distribution, manufacture, and meaning of copper-base metal spirals and hoops through late sixteenth-century and early seventeenth-century northeastern North America. Testing the European derivation of the finished objects and the materials, they use evidence from metallography and compositional analysis to demonstrate that both types of objects were actually made by native people of European-derived materials (smelted copper and brass). They (1991:15) argue quite effectively that native metalworkers used techniques "well within the documented abilities of native Americans." Further, they determine that these objects were fundamentally an Iroquoian trait that became more broadly, but thinly distributed during the early seventeenth century. These objects decreased in frequency as other types of copper and brass items increased.

Importantly, they (Bradley and Childs 1991:16) hypothesize that the meaning of these objects may have had to do with their formal and material connections with native beliefs about the properties and value of copper-base metal materials. For native people, copper meant healing, and healing was the domain of the Underwater Panther in native Great Lakes Algonquian cosmology. The panther, a catlike creature with a long, spiral-shaped tail, was often described as being covered with copper scales. Thus, copper carried great meaning as a healing agent and those who possessed copper possessed this power. The healing power of copper combined with the formal attributes of the spirals and hoops so emblematic of the Underwater Panther may have represented an attempt among Iroquoian peoples to obtain protection from the problem of disease so prevalent among them.

ELUCIDATING TECHNOLOGICAL STYLE IN CONTEXTS OF EARLY MATERIAL ENTANGLEMENT

Bringing Kingery's and Lechtman's approaches together provides a holistic view of technological activity in which both the material and nonmaterial cultural patterns that underpin, sustain, and potentially transform technological and social life are revealed. It is not limited to applying the concept of "choice" in terms of "problem-solving to meet certain technological goals" (after Branstner 1992: 187) nor does it focus simply on producing a usable finished object. The goal here is broader than merely uncovering the technical and cognitive aspects of the manufacturing sequence, as has largely been the focus of Lemonnier's pioneering work (1986, 1993) and has been so evident in some *chaîne opératoire* studies (see Schlanger 1994 for a discussion). While a number of valuable insights into the role of technology and technological change in social life and the importance of understanding production sequences converge in Lemonnier's and Lechtman's work, a critical aspect of the latter's approach is that while she too focuses on uncovering manufacturing histories and processes, she does not stop there (see Hegmon's [1998] discussion of these convergences as well as Dobres and Hoffman's [1994] comparative treatment of their respective positions). She, and others, realize the importance of embedding what is learned about the technical acts of converting a raw material into a finished product within a broader scheme of technological endeavor where it becomes only one (albeit important) aspect of a larger behavioral and symbolic system of "choice" and "action" in material procurement, transformation, use, and meaning (expression) within a particular historic setting.

Importantly for this research, this combined approach is operationalized within a context of understanding the everyday microprocesses of technological activity and change that take place at the level of individual cultures and in spheres of what Peter S. Wells (1991) has called "local meaning." This is significant because during the early phases of culture contact "local meaning(s)" were increasingly under siege first from indirect, then later from direct foreign material, sociopolitical, and microbial influence. As we have seen, identifying and interpreting the multiple meanings and activities associated with increasing appropriation of European-derived objects, particularly of copper-base metals and metal objects, is no simple matter because the transcultural life histories of manufacture, use, reuse, even discard, of these materials is so nuanced and complex.

Combining Kingery's definition of technology as an integrated "system" with Lechtman's "technological style" approach to material culture analysis provides the structure for a rich elucidation of those life histories. Ideally, one would hope that the complete story could be revealed fully; or that we could, as Henry Glassie has put it, truly "locate a genuinely alien consciousness" (Glassie 1988:80). However, as archaeologists working in the native North American protohistoric/early contact period have long known, there is no way to reconstruct and to explain thoroughly every aspect of native acceptance and use of copper-base metals, especially in the earliest, most elusive phases of interaction. However, a materials science approach to artifact manufacture and material composition sheds light on the dynamics of material distribution and native manipulation, use, and reuse that are not discernible through stylistic or formal analyses alone. Also, analysis of protohistoric material culture benefits from the availability of written records. Primary documents, while infrequent, incomplete, and often inherently biased, provide a valuable means through which material change can be interpreted. Taken together, these lines of evidence have unparalleled potential to yield important insights into the technological systems of native peoples and to identify the motives, activities, and processes driving technological change at the level of individual cultures as native entanglement with Europeans intensified.

3 / Recovering Illinois Copper-Base Metalworking Style

The Analytical Program

RESEARCH GOALS

Discovering how the protohistoric Illinois reacted sociotechnologically to the availability of European derived copper-base metals sheds important light on how, why, and in what particular contexts they (as human actors) consumed these materials even as they were becoming increasingly entangled in an ever-expanding commercial and political world systems arena. It brings into sharper focus potentially unique arenas and conditions of initial encounter with foreign materials, which, in turn, illuminate the complex relations among native technologies, material and technological change, and social transformation in the very earliest years of native/European influence.

As outlined in the last chapter, the kinds of responses sought here are constructively understood within a broadened conceptual view of technology and technological activity as a comprehensive *system.* They are particularly well elucidated within an investigative framework of technological "style" that has contextual,[1] spatial, temporal, interpretive, and comparative dimensions. It incorporates the material, technical, behavioral, sociopolitical, economic, and symbolic aspects of artifact manufacture and use. Making such an inclusive investigation truly productive requires developing and implementing an analytical program that not only involves a materials-based technological study of the artifacts themselves, but also incorporates important evidence from the archaeological, archaeometric, and ethnohistorical records. When these multiple lines of evidence converge, we are presented with a broadened database from which to interpret as richly as possible native Illinois metalworking and the social and ideological contexts and implications of European-introduced metals use in Illinois cultural systems. In addition, these findings are then applicable to questions of continuity and change in metalworking practices over the *longue durée* of native copper use in native northeastern North America.

WHY COPPER-BASE METAL?

A limited range of items manufactured of European-derived glass, metal, and textiles are known to have circulated among native peoples of the interior at the time of this study (R. White 1991). Copper-base metals are singled out for study here for a number of important reasons. First, from an archaeological viewpoint, copper-base metals and metal objects are often the first (and only) material indicators of European influence found on protohistoric sites of the period in the Northeastern Woodlands. Second, these materials, consisting of smelted copper and one of its alloys, brass, are quite frequently found on native sites in various stages of reworking. Aside from a few excellent studies, this situation has not received significant, in-depth analytical attention by archaeologists. Third, copper in its native form is known to have been used by some, but certainly not by all, native peoples of the Northeast to varying degrees over millennia. In the protohistoric period, there have been instances in which it has been recovered archaeologically at the same site, in the same depositional context, and in like forms as European-derived copper-base metal (Abel, Burke, and Stothers 2000). Conversely, European-introduced copper-base metal is known to have been widely adopted in cultures that had no deep prehistoric tradition of its use. Its adoption in smelted form, then, begs inquiry into continuity and change in native metalworking practices and contexts of use.

RESEARCH STRATEGY/ANALYTICAL PROGRAM

By virtue of the fact that this study is set in protohistoric/early contact time when written documents (however scanty) are available from at least one of the primary players, an integrated analytical strategy combining the perspectives and the methods of both prehistoric and historic research is adopted here (after Lightfoot 1995). This multidisciplinary approach involves careful, critical integration of evidence from ethnohistorical sources and written documents into archaeological analysis. While the ways in which researchers have chosen to handle these data sources have differed, this combined investigative methodology has become commonplace, even indispensable, for historical archaeologists in material culture studies of early colonization and historic period(s) (see Armstrong 2001; Cleland 2001; Dymond 1974; Hodder 1986; Little 1992; South 1977; Wilson and Rogers 1993).

Numerous authors have pointed specifically to the value of employing such an approach in protohistoric/early contact native contact culture change research (Deagan 1988:8; Galloway 1991; Lightfoot 1995; Perttula 1991; Trigger 1985; Wilson and Rogers 1993). It is utilized here in order to understand the evolution of copper working practices and to contextualize, assess, and interpret Illinois responses to European-derived copper-base metals. The archaeological record and scholarly research resulting from it provide material, temporal, and contextual information about the ways in which native copper was used by prehistoric peoples of the midcontinent. Archaeological data from investigations at the Iliniwek Village supply the material culture context and the copper-base metal sample used in this study. While scarce and frustratingly incomplete, documentary and ethnohistoric records do exist that relate the Illinois to the site and describe their active involvement with foreigners and their merchandise. These sources contribute valuable contextual and historic insights into attitudes, motivations, and actions of both Europeans and native peoples not retrievable from the archaeological record.

However, in addition to these strategies, this work adds yet another crucial line of evidence to the analytical program, that of material science. Heather Lechtman (1977, 1994) has long advocated the use of laboratory techniques as an important, indispensable aspect of materials analysis, especially that of metals, because certain types of laboratory investigations have significant potential to reveal much more about manufacturing and use history and material composition of objects than can be demonstrated by formal or stylistic analysis alone (see also Ciliberto 2000:4; Kingery 1996a, b). Appropriately chosen, these methods provide valuable insights that speak to the technological and cultural behaviors, choices, motives, and expressions of the artisans and of the users that might well otherwise remain outside of our analytical and interpretive grasp.

Rita P. Wright (1989:269–271) provides the scheme for conducting the archaeometric analyses of materials adopted here. Her approach begins with hand examination and low-power microscopy of a large number of samples, then moves to selection of a smaller subsample for greater in-depth examination using and even combining various laboratory techniques such as neutron activation analysis, xeroradiography, and/or scanning electron microscopy. She asserts that such an approach to materials analysis provides researchers with new types of information about the materials, their properties, and their manipulation at more sophisticated levels of archaeometric analysis. This kind of integrated program draws important new analytical methods into the investigative resource base that allow for the generation and testing of new hypotheses related to many larger archaeological questions, including (for example) the interactive relations between technology and exchange.

A large representative sample (n = 806) of copper-base metal artifacts excavated from feature contexts at the securely dated and historically documented Iliniwek Village historic site, Clark County, Missouri, forms the material base of this investigation. The analysis unfolds in a nested series of stages, beginning with formal descriptive (typological/classificatory) analysis and low-power microscopic examination of the entire sample. Metallography, proton-induced x-ray emission spectrometry (PIXE), and instrumental neutron activation analysis (INAA) are then applied to a smaller subsample (n = 75) of selected items from the larger data set to characterize both the production technology of transformed copper-base metal artifacts and the compositions and sources (provenance) of the raw materials from which they were made.

Using this converging analytical framework, this study addresses the following research questions:

1) What is the range of copper-base metal artifact types found in the industry? What can be said of them descriptively? How might they be classified formally and functionally?
2) How and to what extent were these materials transformed by the Illinois? How were objects actually fashioned—what technical processes, behaviors, knowledge, skills, and tools went into the manufacture of the artifacts?
3) What can be learned about the copper-base metal materials themselves? What is(are) their source(s) and composition(s)? Were objects made of native copper, European copper, and/or brass? Was there any cultural preference for one material over another, particularly in the manufacture of specific types of objects or for use in specific contexts? Were there any differences in how a particular material was worked?
4) In what contexts were the transformed artifacts actually used? What was(were) the role(s) and meaning(s) of these materials and artifacts in

cultural life? Importantly, how and to what extent did these materials and objects contribute to cultural and material transformation?

5) What can these fabrication and use processes reveal about technological change and the early processes of transition from native metal use to adoption of European-introduced metal? How does the protohistoric Illinois system of copper-base metalworking and use fit within the larger picture of prehistoric native northeastern North American practices and other modes of protohistoric/contact period copper-base metal usage elsewhere in the Northeast?[2] Is there validity to the notion that there was a single pattern or "native style" in metalworking that extended into the historic period (after Rubertone 2001:136)?

The remainder of this chapter introduces the methodologies used in this study and outlines procedures.

RESEARCH METHODS AND PROCEDURES

I. Data Collection and Sample Selection

The Iliniwek Village Site—Sample Provenience and Collection Methods

The sample of over 800 (*n* = 806) copper-base metal artifacts used in this study was derived from the Iliniwek Village historic site (or Haas/Hagerman site), Clark County, Missouri (a detailed discussion of this site is found in chapter 5). As of summer 1996 when the sampling strategy for this project was set up, portions of four areas of the Haas/Hagerman site had been examined archaeologically (Figure 3.1). These are located in areas A1, C, D, and F1 and F2. While a vague and diffuse Woodland component had emerged at various locations in all of these areas, a single, undifferentiated protohistoric component predominated. Artifact and feature recovery across the site was variable. Block excavations carried out in area C uncovered most of one long house (House 1). Three nearby structures, two long houses (Houses 2 and 4), and one small "single-family" dwelling (House 3), were also sampled. These structures were delineated by post mold patterns. Pit features, some overlapping, dotted the landscape both inside and outside of these structures.

Test excavations were also carried out in areas A1 and F1 and F2. Protohistoric contexts were apparent, but recovering intact features was problematic. Pit features were discovered, but post mold patterns were not detected due to stratigraphic problems in area F and historic ground disturbance in area A

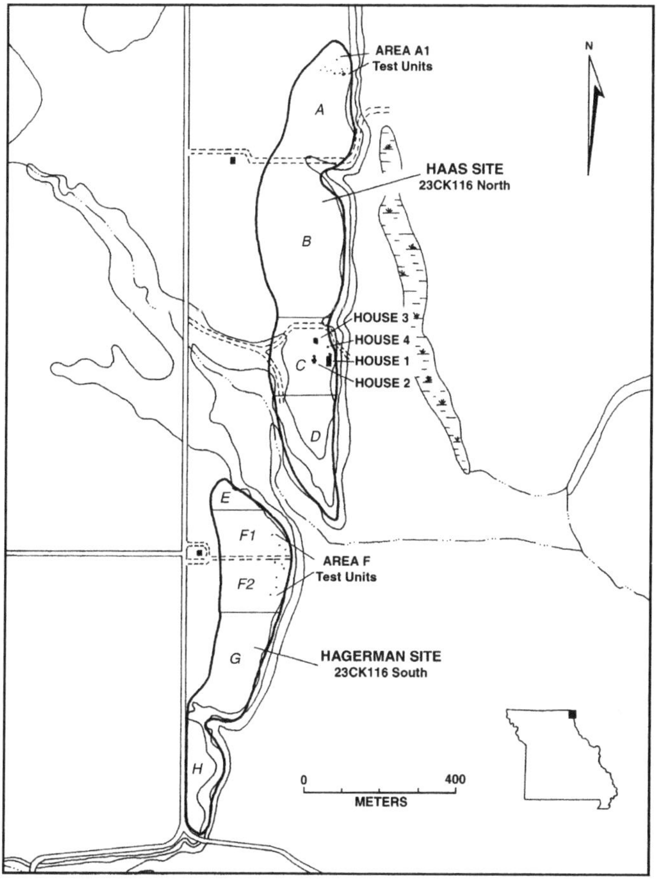

Figure 3.1 The Iliniwek Village (Haas/Hagerman site), Clark County, Missouri, showing area designations and locations of archaeological investigation (after Grantham 1993).

(Ehrhardt 2004). However, based on the similarity in patterning in the contents of the pits in these areas with those in area C, it is likely that they too are associated with domestic contexts. No house floors remain intact in any of the areas investigated, undoubtedly due to farming activities on the terrace during the historic period. Resistivity revealed additional house patterns and associated pits, open areas, and a stockade in area C (Grantham 1996).

Field methods were consistent across all areas of the site. Materials and feature contents were recovered through hand excavation of one-meter squares. Excavation was carried out in 10 cm levels. Features were excavated as autonomous units. In areas A and F, any cultural materials found in features or in matrices thought to be features were piece-plotted to ensure accurate interpretation. All material was passed through 1/8″ hardware cloth. Materials not passing through the screens were returned to the laboratory for further sorting and examination. Selected proveniences in area F were water-screened.

Sampling Strategy and Sample Selection

The sampling strategy was devised so as to obtain a large enough sample to be able to identify the full range of copper-base materials and artifacts in the copper-base metal industry at the site. Once collected, all artifacts in the sample would be classified formally and functionally and then examined technometrically (described below) using metric and low-power microscopic techniques. A selected subsample, again representing all areas of the site, would then be drawn from this large sample for more in-depth archaeometric investigation into manufacturing history and composition.

The 806-artifact copper-base metal sample consists of the copper-base metal contents of 87 proveniences excavated from domestic contexts in areas A1, C, and F1 and F2 of the site (Ehrhardt 2002:97–98, Appendix A). Eighty-one are refuse/storage pits in and around habitation features, and the remaining six are occupation levels associated with domestic contexts. The bulk of the artifacts (n = 737) came from area C, where most intense excavation was carried out. A small but typologically representative sample from areas A1 (n = 15) and F1 and F2 (n = 54) is also included. In all cases, attribution to protohistoric occupation is solid.

The overall sample also includes eight copper-base metal artifacts selected judgmentally both from feature fill and from secure occupational levels in area F1 and F2 for use in a pilot metallography study conducted at the Museum Applied Science Center for Archaeology (MASCA), University of Pennsylvania Museum, in early 1997. The pilot study was undertaken as a preliminary investigation of artifact manufacturing history and composition in order to determine the feasibility of a larger-scale technological study. These artifacts represent both finished and unfinished artifacts and occur in complete and fragmentary form. They were selected because they exhibited surface characteristics reflecting technical processes of native manipulation, reworking, and artifact production such as shearing, folding, bending, rolling (as in shaping), and/or hammering.

Upon determining that the pilot study had yielded interesting and provocative results meriting a larger-scale archaeometric investigation, an additional 67 artifacts were selected judgmentally for that purpose (see Redman

1975:149). Locational context, artifact type, condition, and potential to yield information about technological and use behaviors were major considerations in artifact selection. Emphasis was placed on obtaining multiple samples of most types of finished, unfinished, and partially worked artifact types (when possible) and choosing examples of each that carried visible signatures of manipulation that could be more thoroughly evaluated and interpreted using metallographic methods. The subsample contained artifacts from all three excavated areas of the site (areas A, C, and F).

II. Analytical Methods

Metric/Microscopic Analyses

The purpose of this initial stage of the investigation was to describe the artifacts in the sample formally and functionally and to construct a typology reflecting both their morphological and functional variation and their position in a sequence of artifact manufacture and use. At the same time, artifacts were inspected in order to identify, record, and characterize any attributes that reflected technical manipulation. Many characteristics, such as cutting, scoring, folding, bending, hammering, perforating, riveting, use deformation, and wear were visible to the naked eye. Such signatures occurred on the finished artifacts as well as on those recognized to be partially finished, fragmentary, or abandoned as "wastage" or "scrap" (Franklin, Badone, Gotthardt, and Yorga 1981; Latta, Thibaudeau, and Anselmi 1998).

A program of basic metric analysis and low-power microscopy was followed on all artifacts in the overall sample to identify and describe the range of artifact types in the industry and to record the technical signatures they carried. Jonathan Leader (1988:4) refers to this method of inquiry as "technometric analysis." Length, width, and thickness measurements were recorded for all artifacts. Weight measurements were only recorded for the subsample. This was due to the fact that overall, the artifacts were very small and in many cases, significant amounts of their intact metal had been lost over time. Many occur today in a moderately to highly oxidized state. Consequently, weight measurements were determined to be of questionable analytical utility. Pencil sketches accompanied the descriptive entries.

Along with metric analysis, all artifacts were examined optically with the naked eye and under low-power magnification (6–40x) (Leute 1987:119). This procedure was used in order to detect and to describe in detail signatures of manipulation and use, such as shearing and use wear (Bradley and Childs

1991; Leader 1988:4). An extensive macro- and microphotographic record was amassed to document formal and technological attributes. It is important to point out that during this phase of the examination, no effort was made to clean the artifacts beyond brushing them and/or lightly washing the field sediments off of them. No attempt was made to remove corrosion, as removal of these products (which may be considerable) by abrasion or by chemical means may also remove any surface information the objects might carry. In addition, abrasion may potentially add new striations that may be mistaken for ancient modification (Franklin et al. 1981:20).

A classification coding system was devised and each artifact was placed into a formal/functional/technological category (Ehrhardt 2002:Appendix B). Formal, functional, and technological data from each artifact was then entered onto a spreadsheet. Additionally, each of the artifacts in all typological categories was also classified as to form (finished/unfinished) and condition (complete/fragmentary). Looking at the artifacts in this way allowed for a typology to be constructed that classified the artifacts by form, function, condition, and position in a manufacturing and use sequence. It also allowed for well-informed hypotheses to be generated at this point in the investigation concerning the ways in which the artifacts were manipulated within a proposed production and potential use sequence. These inferences would eventually be investigated further by metallographic methods.

Proton-Induced X-Ray Emission Spectrometry (PIXE)

Modern archaeometallurgical inquiries into ancient technological processes commonly combine some type of compositional analysis with metallography in a complementary analytical program (Wayman 2000; see Bastian 1961 for a discussion of early efforts with native copper), although in some research queries, elemental analysis alone is performed (Hancock 2000:11). The method chosen depends on the questions being asked of the materials and the researcher's flexibility with regard to use of destructive analytical techniques on ancient materials. At the very least, however, archaeometallurgists do wish to *characterize* precisely the material they are dealing with, that is, they want to know of what metal or metals the artifact is made and in what combination. When the behavioral and chemical properties of the metal are known (as revealed by compositional analysis), researchers are better able to evaluate the nature and effects of the stresses applied to specific metals during their manufacturing and use histories, thereby interpreting more accurately the technological meaning of the features revealed in the artifact microstructures (Hosler

1994:18). Then, depending on a number of factors concerning what method was actually used to derive the data, as well as the number and types of elements analyzed and the sensitivity of the technique chosen, resultant chemical "fingerprints" might also be applied to any number of archaeological research questions related to materials sourcing, metal production processes and sequences, artifact manufacture, and distribution, among many others (Hancock 2000).

As one of many types of compositional analysis, PIXE has demonstrated reliability in addressing archaeometric problems of characterization and provenance (sourcing) (Brissaud et al. 1987; Dran, Calligaro, and Salomon 2000; Fleming and Swann 1993; Fleming, Swann, McGovern, and Horne 1990; Hamilton 1996). PIXE is a surface or near-surface method of multielemental analysis. It involves bombarding a material with a charged ion beam. The concentration of elements contained in the material can be determined by analyzing the characteristic x-rays emitted (produced) by the bombardment (Dran et al. 2000:135). After 1970, when true multielemental analysis by energy-dispersive x-ray spectrometry actually became achievable, PIXE gained popularity because it was considered more sensitive than an electron microprobe. Even so, because it is a surface technique, quantification of results is complicated, especially for thick targets when various matrix effects potentially come into play. Software protocols have now taken these effects into account, yielding results within 10 percent accuracy (Dran et al. 2000:135). Concerning its applicability to ancient artifacts, H. K. Gersch, J. D. Robertson, A. G. Henderson, D. Pollack, and C. A. Munson (1998:85) have characterized PIXE as a rapid, nondestructive method with good spatial resolution and moderate sensitivity. Sample preparation is simple and does not require chemical pretreatment.

Although PIXE's application to archaeometallurgical problems appears to be more widespread in Old World research, the technique has been applied in New World copper-base metal inquiries as well. These studies have included characterizing compositionally native copper artifacts from *the* Old Copper Culture (Vernon 1985, 1990) and distinguishing European from native copper-base metals in protohistoric and early contact period contexts (Fleming and Swann 2000; Gersch et al. 1998).

Compositional and sourcing information is essential for researchers studying protohistoric and contact period copper-base metals. It allows them to address directly the single most important question they might ask of the copper-base metal material they recover from protohistoric and early contact archaeological contexts—of what *specific* copper-base metal are the artifacts actually made, native copper, European smelted copper, or brass? These questions are commonly avoided because the presence of corrosion products on the surfaces of artifacts recovered archaeologically precludes making accurate characterization assignments on the basis of visual examination alone. Instead, researchers tend to refrain from distinguishing among the three and refer to all copper-base metal in the industry as "brass" (see Smith 1987:36, and Wray, Sempowski, and Saunders 1991:244 for a few examples). William Fitzgerald and Peter Ramsden (1988) have carried out "scratch tests" of early contact period artifacts (copper "scratches" red and brass "scratches" yellow) to distinguish with the naked eye European-derived copper from brass. However, in separate, subsequent applications, Penelope Drooker (1993), Lisa Anselmi (1994), C. Walker, R. G. V. Hancock, S. Aufreiter, M. A. Latta, and C. Garrad (1999), and this author (Ehrhardt 2002) have come up with inconsistent results when this method was tested against chemical compositional testing.

PIXE is used in this study in tandem with metallography to characterize in a more precise manner the chemical compositions of the metals and alloys under study. In addition, PIXE-derived compositional data is used here to illuminate potential correspondences among material type and source, artifact typology, fabrication, and distribution. What types of copper-base metal materials are represented in the industry? Were the artifacts made of native copper, European-derived copper and/or brass? What are the relationships among material type, artifact type, and fabrication processes? Was a particular metal type chosen for particular types of artifacts? Did native metalworkers work particular materials in specific ways?

Procedures

PIXE was carried out on the entire 75-artifact subset by Dr. Charles P. Swann at the Bartol Research Institute, University of Delaware. The analysis, which involved pointing an ion beam at a raw metal surface exposed on each of the artifacts, was conducted on mounted and ground artifact sections prior to final polishing for metallography. These procedures are described in detail below.

The PIXE procedures followed in this study parallel those commonly used at the Bartol Research Institute for examination of copper-base alloys. The sample is mounted on a stage whereby an area of about 1.5 mm^2 is scanned, giving a good average for the matrix. The proton beam is brought out from the vacuum system of the machine into the atmosphere. Two measurements of the specimen are taken, one to assure that corrosion products (chlorine, potassium, and calcium) are at a minimum. The second is taken using a cobalt filter backed by vanadium, with the beam energy raised to 2.0 Mev. This reduces the intensity of the x-rays from copper downward, and at the same time, allows for a large increase in the proton beam intensity. This permits the observation of x-rays for the elements up through tin and antimony with very good detection limits.

In this case, the samples had already been cut and embedded in preparation for metallography. This allowed for a clean, freshly cut interior section of the artifact to be read by PIXE.

Metallography

Metallography is an invasive technique in which the microstructures of metals and alloys are studied through microscopic examination of polished sections (Chandler 1998:209; Hosler 1994; Vernon 1985). An artifact's microstructure reflects its manufacturing and thermal history, along with something of its composition. Technological interpretations are based on these readings (see Scott 1991:57; Wilson and Sayre 1935). Using metallography, researchers are able to determine whether a metal sample has been cast or wrought (hot or cold worked, drawn), and/or annealed (heated below the melting point). It also provides information as to the degree of cold working and heat treatment a metal object has undergone. Microstructures also reveal what inclusions occur in the metal (Hamilton 1996:13). Hardness tests and various types of chemical elemental analyses often accompany metallographic investigations (see Schroeder and Ruhl 1968; Vernon 1985, 1990).

Along with compositional analysis, metallography is a major component of archaeometallurgical investigations. As Michael Wayman (2000:260) has pointed out, it has been applied to many types of metals for a variety of purposes ranging from gaining a better understanding of past metals technologies to addressing larger-scale historical, archaeological, and anthropological questions related to metals use and human culture (Ehrenreich 1991; Pigott 1991). Cyril Stanley Smith (1981a) has directed attention to its application in conservation and authentication endeavors. Indeed, David Scott (1991:57) has claimed that, "metallography offers one of the most useful means for the examination of ancient metals." Ursula Franklin and colleagues (1981:18) concur, endorsing metallography as "the most informative" approach and the only way in which conclusive proof of the manufacturing history of a copper artifact can be ascertained.

Despite the obvious contributions metallography can and has made to our understanding of many aspects of metalworking behavior, it is underutilized by researchers investigating native North American metalworking technology and contexts of use (but see chapter 4 for important exceptions). This is likely because it is a destructive technique that requires skilled removal of at least one, possibly more than one, appropriate-sized sample from the artifact under study. Once removed, the process involves significant labor-intensive preparation of the surface(s) to be examined. In addition, it also requires that the individual(s) interpreting the resultant microstructure possess specialized metallurgical expertise that an archaeologist normally does not possess. Part of the reason may also be due to the distinct possibility that archaeologists may have once thought that they knew all there was to know about North American metalworking; native peoples handled native copper in a simple and unchanging way over millennia. They never developed techniques that amounted to "real" metallurgy, which involves extracting metals from their ores (smelting) and refining and processing them by such means as alloying and casting (Chandler 1991:243; Vernon 1985:154). Thus, they may have assumed that large-scale, in-depth inquiry into North American techniques would result in a rather limited, redundant repertoire of manipulation techniques including cold working and annealing.

Amidst this apparent intellectual ambivalence, several researchers have expressed interest in reconstructing native metalworking practices using metallographic methods. However, these studies have been limited in scope[3] (see Vernon 1985:156). Even the work of sophisticated historical metallurgists like Smith (1968, 1981a) and archaeological researchers David Schroeder and Katharine Ruhl (1968) on native copper materials, and that of James Bradley and S. Terry Childs (1991) and Helene Dunbar and Ruhl (1974) on contact period copper-base metal artifacts, while illuminating, have been limited to examination of small samples.

Metallography was selected for use in this study for a number of reasons. First, except for Bradley and Childs's (1991) work on early contact period spirals and hoops, Sandra Zacharias's examination of Huron materials (1983), and Eleanora Reber's investigation of protohistoric Madisonville artifacts[4] (1997), little systematic metallographic work had been done on protohistoric artifacts and smelted metals in native hands. This is not to say that these materials had not been studied by archaeologists archaeometrically; it merely means that they had not been studied using metallographic methods. Interpretations of protohistoric and early contact period native metalworking practices had leaned on studies that had been carried out on prehistoric materials. Little attention had been paid to the study of possible change, refinement, or elaboration in native technique or contexts of manufacture or use with the advent of newly available types of smelted and alloyed metal. In this study, it was useful to know whether techniques varied depending on whether metalworkers were using native copper, smelted brass, or smelted copper. I also wanted to test the applicability of Zacharias's hypothesis (1983:9) that protohistoric sheet metalworking "consisted of prehistoric native copper working techniques applied to European materials."

Metallographic analysis was utilized here to determine the manufacturing history, composition, and to a lesser extent, the provenance of a selected number of the artifacts in the subsample. Specifically, it was employed to assist in distinguishing European (smelted) copper from native copper and to characterize Illinois fabrication sequences and metalworking techniques. These questions had been asked of copper-base metals from protohistoric contexts before; however, one of the contributions of this study is that it considers a large sample size and multiple examples from several artifact categories (Bradley and Childs 1991; Wayman, King, and Craddock 1992).

Procedures

Metallographic investigations commonly involve a number of standard preparation and examination protocols that were followed here. These protocols are described in Scott (1991). In brief, these steps involve 1) selection of the sample and area of interest; 2) sectioning; 3) mounting; 4) grinding, polishing, and etching; and 5) examination under the microscope (Chandler 1998:209–212). Proper preparation of the sample is essential, as improperly prepared specimens can result in incorrect interpretations and time-consuming repreparation, not to mention damage to the artifact and/or to the sample (Vander Voort 1999:60).

Except for perforated pendants, multiple examples of each major artifact type identified in the descriptive analysis were included in the metallographic sample, along with single examples of less frequent types, such as rings and spiral beads. The two artifacts suspected of being made of native copper were also included. Each of the selected artifacts also carried one or more indicators of manipulation such as shearing, folding flush, bending, hammering, and/or perforating. These areas of the artifact were targeted for examination. It is important to note that only 64 of the 75 artifacts sectioned and mounted as required for PIXE were ultimately examined metallographically. Because preparation and examination is so time consuming and costly, it was determined that if sufficient redundancy was reached in the results of the metallographic examination of any one type of artifact, no further testing of that artifact type would be pursued.

Each artifact was examined carefully before testing began in order to 1) verify its suitability (in terms of its condition[5]) for analysis; 2) delineate the zone(s) of technological interest; and 3) decide upon the optimal yet least invasive sectioning scheme to obtain the most technological information, especially if more than one area of interest was noted. Pencil drawings were made of each of the artifacts selected with the areas of interest delineated and the direction and extent of the intended sectioning indicated. Before any artifact was cut, it was photographed again for archival purposes, regardless of prior photography.

Metallographic samples were cut from the artifacts and mounted in round plastic molds using a cold mounting process. Once completely hardened, mounted specimens were subjected to a sequence of progressively fine surface polishings in order to achieve what is referred to as the "as polished state." Etching followed with potassium bichromate and ferric chloride, or less frequently with Slepian and Prohaska, Klemm's III, or ammonium hydroxide. Sections were examined under a metallurgical microscope at magnifications ranging from 55x to 600x, first in an "as polished" condition, then in an etched condition. Microphotographs were produced using positive/negative, black-and-white Polaroid photography. At least one microphoto was taken of each sample.

Observations were made across each sample concerning the range, size, and type of grains, the distribution and nature of inclusions and corrosion products, the nature, extent, variability, and heterogeneity of signatures of manufacturing history. These features include but are not limited to the presence of strain lines and twinning within the grains, intergranular or transgranular cracking, internal bursts, as well as grain boundary thickening or other indications of a two-phase structure (after Scott 1991:67).

Instrumental Neutron Activation Analysis

Instrumental neutron activation analysis (INAA) is another chemical compositional technique that has been applied widely throughout prehistoric native

North America to problems of provenance in pyrotechnical materials, particularly ceramics and native copper (Goad and Noakes 1978; Hancock, Pavlish, Farquhar, Salloum, Fox, and Wilson 1991; Kuleff and Pernicka 1995; Levine 1996; Neff 2000; Rapp, Allert, Vitali, Jing, and Henrickson 2000; Wayman, King, and Craddock 1992). INAA is a bulk analytical technique of high precision, accuracy, and sensitivity; sensitivities can reach the order of parts per billion or better. The technique is relatively free of matrix effects. For these and other reasons, it is often considered the "technique of choice" in provenance studies (Glascock 1992; Glascock, Neff, Cogswell, and Herrera 1996:7; Neff 2000:102–103). Its major disadvantage in studies of copper-base alloy characterization is its inability to measure lead (Pb), which is often added to these metals to improve workability (Kuleff and Pernicka 1995:159).

Unlike PIXE, however, INAA is a destructive technique that *requires* removal of a sample of material to be tested. While only small samples are needed, INAA also requires careful sample preparation to minimize contamination and weighing error. In addition, upon testing, INAA samples are rendered radioactive, requiring special storage and disposal (Neff 2000).

Both Michael Glascock (1992; Glascock et al. 1996) and Hector Neff (2000) have explained how INAA works. This will not be repeated in detail here except to say briefly that when neutrons bombard a sample, a small fraction of the nuclei of each of the constituent elements becomes excited and transforms into unstable radioactive isotopes that de-excite (decay) according to their respective characteristic half-lives. These isotopes emit gamma rays whose energies are also characteristic of the elements. These gamma rays can be measured, yielding a determination of the quantities of various elements in the sample (after Glascock 1992:12). INAA typically measures 20–40 elements.

Rooted in the work of Sharon Goad and John Noakes (1978) and George Rapp, James Allert, Vanda Vitali, Zhichun Jing, and Eiler Henrickson (2000) on sourcing native copper and copper artifacts through prehistory, INAA has emerged as an important, effective technique for addressing problems of composition and sourcing native and European-introduced copper-base metal materials from protohistoric and early contact period sites in North America. Canadian researchers have been at the forefront of this research. They have demonstrated that the technique is useful for characterizing and distinguishing with confidence European-introduced and native copper-base metals (Wayman, Smith, Hickey, and Duke 1985). Studies carried out by Hancock and various collaborators at the now-decommissioned SLOWPOKE Reactor at the University of Toronto have also addressed problems of sourcing, chronology, and trade in European-introduced copper-base metals from protohistoric contexts at Madisonville (Ohio) (Drooker 1993, 1996b) and in Ontario (Hancock et al. 1991; Hancock, Fox, Conway, and Pavlish 1993; Hancock, Pavlish, Farquhar, Julig, and Fox 1994; Hancock, Pavlish, Fox, and Latta 1995; Moreau and Hancock 1999; Walker, Hancock, Aufreiter, Latta, and Garrad 1999). Another related application of INAA "fingerprinting" involves attempts to "group" similar individual chemistries of European-introduced copper and brass to identify and isolate "lots" or "batches" of kettles. These studies are promising for exploring intrasite distribution and consumption of trade kettles as well as patterns of interregional trade (Anselmi, Latta, and Hancock 1997).

INAA was employed here as a complementary compositional technique to PIXE. It identifies a different suite of elements and has greater precision than PIXE in chemical fingerprinting due to its decreased susceptibility to matrix influences (see Gersch, Robertson, Henderson, Pollack, and Munson 1998 for a practical example). Individual chemistries of the 75-artifact subsample were sought in order to make inferences about the number and variety of trade metal (kettle) sources represented at the site, how fragments from the same kettles or "batches" of kettles might be distributed across the site, whether there were preferences in material selection for particular objects, and what the results could contribute concerning the sequences and contexts of modification of individual copper-base metal kettles.

Procedures

Because it is a destructive technique, INAA followed metallography and PIXE in the analytical program. For that reason, and so that results achieved from PIXE and INAA might eventually be compared, the study required taking samples from the same artifacts used for PIXE and metallography.[6] The study was carried out at the Missouri University Research Reactor (MURR). Their protocols for sample preparation, irradiation, and analysis were followed. A short synopsis follows.

Preparation

A minimum sample weight of 15 mg was needed for irradiation. New samples were either cut from the artifacts or samples used for metallography were removed from their molds for use in INAA. To minimize contamination, removal of surface and penetrating corrosion products is strongly recommended prior to irradiation. Corrosion products were cleaned by immersing the samples in a small amount (< 5 ml) of 2.5M HNO_3 (nitric acid) then rubbing them with cotton swabs. Native copper required less than five minutes in the solvent, while European-derived copper cleaned rather well after one hour. Brass required at least 1–1 1/2 hours in solution, and yielded the least satisfactory result. Duplicate samples were taken from four artifacts but were not cleaned. This resulted in a 77-sample data set.

Samples (unknowns) were weighed into labeled 4 × 6 mm quartz vials and their tops sealed. They were then bundled for irradiation according to the amount of zinc they contained (Bundle 1 = copper; Bundle 2 = brass [containing over 2 percent zinc by PIXE]). Two sets of standards consisting of unalloyed copper (SRM400), silicon bronze (SRM158a), cobalt-aluminum wire (.116 percent cobalt), and gold-aluminum wire (.058 percent gold) were also prepared, weighed into vials, and labeled. A set of standards was packaged with each bundle.

Data Collection/Irradiation

Bundles were irradiated at MURR for four hours in a neutron flux of 5×10^{13} neutrons per square centimeter per second. After a decay of seven days, samples were counted on a high purity germanium detector for 30 minutes to measure gamma rays emitted for short-lived elements copper, arsenic, and gold. Three weeks later, samples were counted a second time for two hours on the same detector to measure long-lived elements silver, cobalt, iron, nickel, antimony, selenium, and zinc. Absolute concentrations were determined relative to the known standards.

In the first irradiation and counting process, ^{64}Cu (copper) isotope was measured to obtain copper concentrations. Since it is a positron emitter, interference problems resulted that produced spurious copper values. To correct the problem, a second irradiation and count was conducted using new samples from 10 of the specimens in the first data set. The values produced were used to calculate a correction factor that was used to adjust the ^{64}Cu-based values to ^{66}Cu-based values. ^{66}Cu is a shorter-lived isotope that causes less interference. The counts on the other elements in the data set were not affected.

Data Analysis

Compositional data was examined using multivariate pattern recognition and statistical techniques. At MURR, pattern recognition is achieved through principal components analysis (PCA), a powerful technique whose advantages in provenance research over both cluster analyses and bivariate scatterplots are explained by Neff and Glascock (1995:281–282) and will not be restated here. PCA is used to define initial compositional groups and to identify subgroups within these groups. Group membership and the cohesiveness of suggested groups may then be demonstrated using bivariate plots; however, MURR also evaluates group membership statistically by using Mahalanobis distance calculations. Again, the rationale for use of this technique will not be discussed here, except to say that it has demonstrated efficacy in assessing group membership recognized by multivariate statistical means.

Historical/Documentary Research

As discussed at the beginning of this chapter, archaeologists studying contact period change and continuity routinely do so from an interdisciplinary approach, which, according to Mark Leone and Parker Potter (1988:14) "marries" historical and archaeological investigative strategies. According to Samuel Wilson and J. Daniel Rogers (1993:6), "archaeological information is essential in providing an extended diachronic dimension to ethnographic and ethnohistorical studies, and in some cases it offers the only data available on the precontact characteristics of Native American societies." At the same time, they also recognize the ways in which ethnohistoric information can benefit archaeological research of the period. They (1993:7) state that for archaeologists, in making "substantial use of the documentary record . . . sociocultural realms of meaning loom larger in their explanatory potential than in corresponding interpretations of prehistoric periods." Importantly for this work, they argue that, "the contact period offers an opportunity to explore the *meaning* of artifacts and the relationship between material goods and cultural categories [emphasis theirs]."

To achieve these goals, Leone and Potter (1988) suggest that the archaeological record and the documentary record be considered two independent sources of evidence about the past, and that investigators "work back and forth, from one to the other, using each to extend the meaning of the other." In this way, the "internal logic" of each source is maintained, and neither historical nor archaeological evidence becomes background for the other (Leone and Potter 1988:17). In this research, both are used in a truly complementary way to enrich interpretations of the motives for and meanings of the technological shifts revealed in the material record and attested to in its documentary counterpart.

Written documents[7] in many forms were examined for what they might reveal about 1) the ethnohistoric Illinois and their sociopolitical relations with other native peoples and Europeans at contact; 2) past copper-base metalworking technologies and the significance of copper to native peoples, including the Illinois; and 3) the activities and interests of Europeans as they expand into the interior. Contemporary (primary) sources of evidence, mainly in the form of eyewitness travel and exploration accounts, maps, and written communications between New World satellites and their European headquarters were mined to tease out whatever statements may be germane to the above questions (see Sayre 1997). Secondary historical, anthropological, and archaeological sources that interpret events, activities, and processes of change were also consulted as was modern ethnohistorical scholarship. While early manuscripts and the statements contained within them had to be carefully assessed for reliability and inherent biases, they remained an important line of evidence from which to draw inferences.

This integrated research program reveals the Illinois' copper-base metalworking "style" in multiple contexts and dimensions. While documentary and archaeological evidence have become part and parcel of historical archaeologists' analytical repertoire, pulling in laboratory and materials science approaches provides a window into technological response unavailable through other means. Bringing these lines of evidence together in this way brings into focus the choices, behaviors, and processes of social action involved in understanding native consumption of foreign goods and merchandise. The remainder of this work makes these activities explicit, revealing the microenvironments of change among the Illinois during the earliest pulses of European expansion into the Western Great Lakes.

4 / Indigenous Copper Working in the Midcontinent

Situating Illinois Copper-Base Metal Use in Late Protohistory

In any investigation of indigenous copper-base metal use in early contact times, it is important to keep in mind that native peoples' exposure to European-derived metal does not constitute their first encounter with copper as a raw material. On the contrary, it is simply a new phase in a long tradition of indigenous copper metalworking. Copper had been exploited by indigenous peoples of northeastern North America in its metallic, or *native* form for nearly 7,000 years before Europeans ever introduced them to it in smelted and alloyed form (Martin 1999:143). Over that long period, its consumption by prehistoric groups was hardly regular or ubiquitous. However, even among cultures that used it extensively, the technology never reached the level of sophistication recorded for Central and South American traditions. Conversely, some groups who lived near native copper sources apparently never exploited them (Kraft 1996:33). To further complicate the picture, when European-introduced copper became available, cultures with no strong metalworking traditions adopted it and used it in creative ways (Childs 1994:229; Rubertone 2001:136–138).

In this chapter, protohistoric copper working and use in the midcontinent is placed in technological and historical perspective over the *longue durée* of prehistory, protohistory, and early history (Cleland 2001:3; Lightfoot 1995: 202). This type of information forms a critical baseline from which arenas of stability and transformation in protohistoric/early historic period metal-working and use systems can be identified. We are then in a position to assess the degree to which materials choice, techniques, and/or contexts of use in the protohistoric mirror the prehistoric past or represent important new technological, stylistic, and symbolic trends (after Lightfoot 1995:205).

An exhaustive treatment of copper as a raw material, its manipulation and use by native prehistoric and early historic cultures of the Western Great Lakes and North American midcontinent, and the techniques applied to its study, is quite impossible here.[1] Instead, this review centers on three major prehistoric copper working traditions of the midcontinent—*the* Old Copper Culture (Archaic), Scioto and Havana Hopewell (Middle Woodland), and Mississippian. The overview continues into the early contact period, focusing on what is known about native metalworking techniques using European-derived copper-base metals. It ends with a discussion of the symbolic meanings copper held for native people.

COPPER—ITS PROPERTIES, SOURCES, PRODUCTION, AND FABRICATION METHODS

Physical Properties of Copper

Copper is a nonferrous metal with a face-centered cubic crystal structure (FCC). It is reddish to yellowish red in color and lustrous when polished. It has been known for thousands of years, and was probably the first metal to be used by humans (Brady, Clauser, and Vaccari 1997:248; Craddock 1995: 97; Kuleff and Pernicka 1995:145; Maddin, Wheeler, and Muhly 1980:211; McCreight 1991:8; Smith 1981a:77; Wayman 1989:3). It is resistant to corrosion and is an excellent conductor of electricity (Chandler 1998:58; McCreight 1991:8; Scott 1991:139). Copper is a relatively soft metal (approximately Rockwell F. 40) with excellent working qualities (Hodges 1976:64; Untracht 1968: 17). It is highly ductile and has good tensile strength. It is easily fabricated and can be formed hot or cold. It has good joining and soldering characteristics. As such, it is one of the most important and widely used metals (Untracht 1968:17).

Native Copper Sources and Early Production Techniques

Copper Sources

Copper occurs in native form and in a large number of ores. In its native, or metallic state, it is found in large dendritic masses or massive chunks, in smaller lumps, in thin plates, or in arborescent forms (Maddin et al. 1980:212). Native copper is generally known to be quite free of impurities (usually in excess of 99.9 percent pure), however, the amount and pattern of impurities can vary depending on the copper's geological source (Maddin et al. 1980:214). Trace element impurities typically include silver, arsenic, and iron, with lesser amounts of nickel, tin, chromium, indium, and antimony (Wayman 1989:4).

Copper is the most abundant of the native metals and is found on every continent inhabited by early metalworking cultures (Wayman 1989:3). In North America, the most significant deposits occur in the basaltic lavas of the Lake Superior region (Craddock 1995:94–95). Centered on the Keweenaw Peninsula and Isle Royale, Michigan, locales, outcrop, near surface, and vein deposits occur. These are known to have been heavily mined by native people since antiquity (Griffin 1961). Remains of extensive surface extraction are evidenced by the irregularly sized and shaped pit and trench surface mines dotting the landscape (Martin 1999:85, 90). "Float copper," that is, nuggets or pieces removed from these original deposits by glacial action, was also available to native people in stream beds or other depositional contexts hundreds of miles from their points of origin (Wayman 1989:3).

According to Michael Wayman (1989:4), prehistoric mining operations were carried out by setting fires directly on the deposit surface, then sweeping them away or quickly quenching them with water when sufficiently high temperatures were reached to fracture the host rock. Copper was then extracted from these fractured pieces. It was further worked with hammerstones, mauls, chisels, and wedges, which were often left in the pits created by its removal. Sometimes large or unwieldy pieces of copper are found abandoned in ancient mining pits, unable to be removed.

Because of its abundant, well-known, and highly visible copper resources, many archaeologists have long believed that the Lake Superior district was the central, perhaps the only important source of native supply. However, archaeologists have recognized that additional sources may have been exploited as well (Levine 1999). Because it bears so directly on questions of distribution and trade, determining the raw material source or sources for the copper artifacts found in the prehistoric Eastern Woodlands has long been of interest. Early scientific investigators included Clarence Moore (1903), Curtis Wilson and Melville Sayre (1935) and Roy Drier (1961). For the last 25 years, important geochemical research has been underway to test that hypothesis scientifically and to link artifacts reliably to their sources. Using neutron activation analysis (NAA) to detect trace element patterns in samples from over 65 source localities, George Rapp et al. (2000) have made significant inroads into chemically fingerprinting North American copper deposits and linking them compositionally to prehistoric copper artifacts. Seventeen separate source locations have been defined with confidence, seven of them from within the Lake Superior region (Rapp et al. 2000:88–89). They (2000) have also tied 21 native copper artifacts from Minnesota to four different Lake Superior region sources.

Using the same principle and analytical technique with artifacts and copper samples from farther afield, Mary Ann Levine (1996) identified twelve discrete geological sources of native copper, five from east of the Lake Superior district (Levine 1996:124). Her linkages between the 64 artifacts she tested and sources identified in the combined Rapp/Levine database remain inconclusive, however, because of small sample size/small database and overlaps in trace element signatures (Levine 1996:171, 197). Much is learned from these types of studies, and despite these difficulties, the database for this type of analysis continues to grow and the method continues to hold promise.

Copper Smelting

Copper is known to be separated easily from its ores. This pyrometallurgical process, known as smelting, involves extraction of primary metals from ores through the use of high heat in a reducing atmosphere by burning of a fuel (Untracht 1968:3). Copper is thought to be the first metal smelted. Smelting operations have been known as far back as the sixth millennium B.C. in the Old World but became more common in the fourth millennium (Craddock 1995:126). While there is no tangible evidence for the beginnings of this process, metallurgists have speculated on its origins. Paul Craddock (1995:122–123) posits that it is thought to have progressed from heat treatment of native copper. Use of heat treatment led to melting, then progressively to smelting operations. Henry Hodges (1976:68) conjectures that it happened first when metal was inadvertently smelted in a campfire or in a kiln. Neither melting nor smelting processes ever developed in prehistoric northeastern North America.

Copper Alloying

Copper also combines with many other elements and serves as the base metal for a significant number of wrought (worked) and cast alloys, or admixtures of two elements, the principal one being a metal (Brady et al. 1997:31). Today, several hundred wrought and cast alloys are available (Brady et al. 1997:251; McCreight 1991:8). Alloying copper alters its properties. It can make it harder and more suitable for manufacture of tools and weapons (Hodges 1976:64). For instance, adding lead in small proportions (to 10 percent) increases the fluidity of the metal and assists in casting; adding antimony and arsenic (about 3 percent) hardens the copper.

Early metalworkers in both the Old and New Worlds alloyed copper. The earliest alloy was arsenical copper. Other early, major alloys are bronze and brass. Bronze, a binary alloy of copper and tin, appears in antiquity in the Old World, replacing arsenical copper over time. Brass, another binary alloy consisting of copper and zinc, appears much later there. Early ternary alloys (containing three major constituents) also occurred (Hodges 1976:68–69). Unlike Central and South American metalworking traditions in which binary and ternary alloying was practiced (Hosler 1994; Lechtman 1977, 1994), alloying was not known in native northeastern North American copper working traditions. The high arsenic content (4.14 percent) of a Hopewellian copper ear spool found in a British Museum collection is thought to have been a natural occurrence of the raw material potentially chosen by metalworkers for its color (Wayman, King, and Craddock 1992:100).

The European brass that became available to native peoples in the contact period ranged in zinc content from about 2 percent to somewhat over 30 percent. During this period, zinc is thought to have been added to copper through the cementation process wherein copper metal is mixed with zinc oxide or carbonate (calamine) and heated in a closed crucible. Using carefully regulated temperature, zinc ore is reduced to metallic zinc vapor, which diffuses into the copper. This is followed by melting the metal under increased temperature to homogenize it (Bayley 1998:9).[2] Smaller amounts of other elements such as tin or lead may have been added. However, these additions may have compromised the effectiveness of the cementation process (Bayley 1998:9). They also change the working properties of the metal (Brady, Clauser, and Vaccari 1997:119–120).

Generally, brass has good mechanical properties, good corrosion resistance, but lower electrical conductivity than copper. Like copper, its strength and ductility depend on alloying and/or cold work. In modern industrial terminology, copper alloys with zinc content in copper to about 30 percent are termed *alpha brasses.* Within this category, a wide range of zinc contents is seen in imported copper-base metals, ranging from gilding metal (5 percent Zn) to cartridge brass (70 percent Cu/30 percent Zn) (Brady, Clauser, and Vaccari 1997:121).

Copper Fabrication (Shaping) Techniques

Once copper is extracted (and perhaps smelted and/or alloyed), the metal is fabricated, producing a desired finished shape. In antiquity, copper was fabricated in one of two ways, casting or working (forging), including drawing (Hamilton 1996:15). In many cases, both processes were involved in the manufacture of a finished product; casting was the primary process and working followed. Casting is an early fabrication process wherein totally molten metal is poured into a mold cavity of a desired shape. Upon solidification, the metal assumes the shape of the mold (Callister 2000:354). Archaeometallurgists widely hold the position that indigenous northeastern North American metalworkers did not melt copper, thus, they did not use casting methods to fabricate finished objects.

Rather, native metalworkers formed copper by working (forging) it into shape. Forging operations can be carried out with the metal at a temperature above which it recrystallizes (*hot* working), however, overwhelmingly, native northeastern North American copper workers *cold* worked native copper into final shape through hammering procedures that deformed the metal by compressing and spreading it (Callister 2000:353; Martin 1999:117). This means that artifacts were shaped at ambient temperatures through hammering, probably with cobblestones, and bending[3] operations. During this process the metal deforms plastically and its grain structure becomes increasingly deformed, producing what is called a "work or strain hardened" effect. Extremely strain hardened copper eventually becomes overworked, or embrittled. The effects of strain hardening and embrittlement can be reversed by heating (called annealing) the object to a temperature as low as 120° C to over 800° C but below its melting point (1085° C). Through annealing, copper deformed by cold working recrystallizes. It becomes soft and regains its ductility and malleability (Vernon 1990:501–512). Rate of recrystallization depends on the thickness of and impurities in the material, the temperature and duration of the anneal, and the degree of prior cold working. In executing finished artifacts, cold working and annealing cycles can and probably were repeated as needed until the finished form was achieved. However, David Schroeder and Katharine Ruhl (1968:162) have argued that the malleability of native copper obviated the need to recrystallize the metal's structure in order to work it heavily.

INDIGENOUS METALWORKING PRACTICES IN THE NORTH AMERICAN MIDCONTINENT—A REVIEW

The Old Copper Culture (3000–1500 B.C.)

It is probably no accident that the earliest major copper working cultures appear near the copper-bearing zones of Isle Royale and the Keweenaw Peninsula. Throughout the Archaic period (5–6000–1000 B.C.) numerous distinctive prehistoric cultures of the Western Great Lakes region exploited this unique geophysical landscape. While some of these groups have erroneously been lumped together as "*the* Old Copper Culture" after their shared copper working technology, Guy Gibbon (1998:40) correctly prefers to call this phenomenon a "metallurgical tradition shared in varying degrees by a number of different societies." It is most densely evidenced in eastern Wisconsin, but is widespread throughout the region and well beyond (Halsey 1996:11; Mason 1981:181). This metalworking tradition is characterized by the production of copper implements, weaponry, and ornaments. While found in domestic spheres, objects are featured mainly in mortuary contexts, and are often found in large caches (Halsey 1996:11; Mason 1981:181). The industry includes various types of implements and ornaments, such as socketed or tanged spear points, knives, straight, or "crescent" type blades (ulus), awls, perforators, spuds, celts, chisels/wedges/axes (some also socketed), fish hooks, gorgets, crescent–shaped ornaments, bracelets, and beads. Artifacts tended to be heavy, bulky, and primarily utilitarian in form (Leader 1988:73). While knives are largely thought to have been undecorated, Christopher Stevenson (1976:138) has argued that a few knives appear to be embellished with drilled or punched depressions placed in paired or in linear fashion along their blades. One adze carried paired linear zigzags along the central axis of the blade.

While formal, metric, and/or low-power microscopic analyses have been employed to classify and describe these materials, sophisticated laboratory techniques have also been called upon in order to learn more about Old Copper metalworking technologies. Xeroradiography and metallography are the principle methods chosen. A noninvasive technique, xeroradiography (Leader 1988:5–6) allows researchers to identify macro-attributes that reveal fabrication techniques, in situ contact between metal artifacts (in some cases), and evidence of repair and reconstruction. Its main advantage is that large numbers of whole artifacts can be examined in a nondestructive manner as opposed to the time-consuming spot sampling, preparation, and examination of small numbers of objects as is typically the case with metallography.

Metallography has also been performed on Old Copper materials (Schroeder and Ruhl 1968; Smith 1968; Vernon 1985, 1990). Aimed largely at revealing Old Copper manufacturing processes and production sequences, these studies reveal that artifacts of the period were fabricated mainly from nuggets or laminar plates of copper that were hammered, flattened, and repeatedly folded over and bent into shape. Alternate cold working and annealing processes were employed in manufacture. Varying degrees of cold working and annealing were accompanied by differential, often combined use of abrading (grinding), riveting, perforating (by drilling), bending, or molding (often around mandrels), and polishing techniques (Leader 1988:71; Vernon 1985, 1990; see also Martin 1999). Researchers have found no evidence of casting, melting, or smelting the metal (Smith 1968:242).

The extent to which hot working (forging) techniques were utilized by Old Copper artisans has been difficult to determine (Drier 1961; Martin 1999). In this working method, the raw material is shaped at a rate and temperature at which strain hardening cannot occur. Copper is worked at a white-hot or hot-short (above 1000° C) temperature—just below melting (Martin 1999:132). Jonathan Leader (1988:71) claims this forming technique is a feature of this tradition, but supplies no evidence to support his assertion. In his detailed metallographic study of objects from the University of Pennsylvania Museum, William Vernon (1990:502) also saw no evidence for hot working. However, he points out that one difficulty has been that metallographic analysis is not conclusive in distinguishing hot forging from cold working, since they both result in very similar microstructural features (see also Martin 1999:132). Vernon (1990:502) has also argued that there is no evidence during this period for handling hot metal with tools in a forging process.

The issue of hot working notwithstanding, these results indicate that copper artifacts of the period are generally complex in form and involved significant technical proficiency and knowledge on the part of the artisans of both the behavior of the material and of manufacturing techniques (Leader 1988:71; Schroeder and Ruhl 1968:162). Leader (1988:71) has claimed that some of the artifacts "represent ingenious solutions to basic engineering problems . . . while others may represent solutions to problems that we can only dimly perceive."

One problem remains, however. Ronald Mason (1981:186) has characterized Old Copper Cultures as ones that

> would have shared in a varying degree a common metallurgical technology and a set of style concepts regarding those many potential attributes not directly dictated either by the nature of the material or by the specific *tasks the tools were meant to perform* [emphasis mine].

His statement brings up an important question concerning how these artifacts actually functioned in social life. While he (1981:187) and others have claimed that many of these artifacts were manufactured in the form of implements and were parts of "tool kits," many are found in mortuary contexts, so to what extent they may have actually ever been put to use in a practical (what Binford [1962] has called "technomic") sphere has been an interesting question. Clues have come from visual inspection, wear patterns, and internal characteristics.

Although Leader (1988:50–51) points out that while corrosion generally masks wear patterns, his visual and xeroradiographic research revealed signs of use and resharpening on many objects, including implements from a Sault Ste. Marie, Michigan, cache (whether this was a mortuary cache was not clear). He (1988:62–63) found wear patterns and use damage on square edged celts. Chisels also showed signs of edge chipping and resharpening. Edge damage was also noted on socketed spearpoints. Knives are thought to have been manufactured using special "hammer packing" techniques to enhance the efficacy of their cutting edges (Leader 1988: 51, 55, 65, 68–69). John Penman (1977), who also found use wear on several types of Old Copper tools, argues that among other types, many knives, socketed projectile points, and perforators (awls) were indeed heavily used.

Vernon (1990:510–511) uses microstructural evidence of artifact manufacture to weigh in on the question. He found that three types of artifacts, including points, knives, and celts, were left in an annealed state. These outnumbered those left in a cold worked condition nearly four to one. Awls, however, were found left in a cold worked condition. He reasons that if cold working hardens and strengthens copper artifacts, then we should find those that were used or intended for use left in that state. Artifacts found in an annealed condition would not have necessarily been intended for use because the metal was left in too soft a condition to withstand the strain of use. He argues that since they were left cold worked, awls functioned in a utilitarian sphere, likely because of their "advantage over similar bone or stone awls; that is, they would not break." Remaining objects were likely manufactured for a "special" or nontechnomic function, as burial accouterments, for example.

Researchers do not agree on how to interpret these findings. Vernon maintains the cold worked = utilitarian/annealed = nonutilitarian argument, while Gibbon (1998:40) has said that the working qualities of copper would permit an annealed, then heavily used and deformed blade to be reshaped easily. As Gibbon points out however, the fact that these artifacts have been used does not necessarily preclude them from ending up in burial contexts. In any case, Old Copper artifacts are far from exclusively ideological in function. According to Leader (1988:70–71) it is difficult and probably imprudent to make single-use categorizations.

Scioto and Havana Hopewell Traditions/Middle Woodland Period (200 B.C.–A.D. 400)

Although copper working continues in the Western Great Lakes and is certainly evidenced elsewhere in the Northeast over subsequent centuries (Childs 1994; Halsey 1996:11; Levine 1996), the next notable florescence of copper working technology occurs in the Middle Woodland period (200 B.C.–A.D. 400). As in other regions of the midcontinent and Southeast, elaborate copper objects have been recovered among cultures of the Ohio Valley and the Illinois/Mississippi River valleys. The Scioto Hopewell and Illinois/Havana Hopewell were focal in what is known as the Hopewellian Interaction Sphere (Struever and Houart 1972). Considered an "exotic" raw material, copper was one of the primary commodities circulated interregionally (Goad 1979:239; Seeman 1979:291–292). Copper working activity is reflected in part by partially worked masses or nuggets and incomplete artifacts. Both finished and unfinished artifacts are documented primarily from mortuary contexts (Winters 1981:19). A wide range of primarily ornamental forms, including ear spools, gorgets, bracelets, beads, arm rings/bracelets, and head dresses, is featured in the industry, while far fewer utilitarian implements, such as copper axes, celts, adzes, and gouges, are found. Copper cutouts, panpipes, and platform pipes also occur.

Hopewellian Interaction Sphere copper artifacts, primarily those of the Scioto tradition, have been studied archaeometallurgically by several researchers. In the 1890s, Charles Willoughby (Greber and Ruhl 1989) provided detailed descriptions of many of these artifact types. Both he and Frank Cushing (1894) conducted replicative copper working experiments to try to reconstruct Hopewellian manufacturing processes and to ascertain the capabilities of native metalworkers. Much more recently, important research has been conducted by Schroeder and Ruhl (1968), Leader (1988), N'omi Greber and Ruhl (1989), Wayman, J. C. H. King, and Craddock (1992), and Ruhl and M. F. Seeman (1998). Sharon Goad (1979) and John Noakes (Goad and Noakes 1978) have investigated Hopewellian copper materials from southeastern complexes. Materials analyses range from visual inspection and description to various types of compositional, provenance, and work history characterization, and experimentation. Salient findings are synthesized here.

Among Middle Woodland Ohio Hopewellian peoples, artifacts of utilitarian and ornamental function were manufactured by hammering copper nuggets or prepared blanks into solid forms. Blanks or rods were formed by hammering, annealing, and grinding into shape (Leader 1988:78). Final forms, such as solid bracelets and adze heads, were achieved by cold working and annealing, and various combinations of bending, folding, rolling, and grinding activities (Leader 1988; Wayman et al. 1992:107, 113). Sockets were achieved by hammering over an anvil (Smith 1968:242). Holes were created by punching or drilling. Metallographic analysis (Wayman et al. 1992) revealed that the heads of two solid adzes examined were formed using either hot forging or cold hammering and annealing. In the case of one adze head, the final treatment was cold working, which Wayman and colleagues (1992:112) agree "serve(s) to increase the hardness of the blade, making it more functional." Tentative assessment of another was that it was left in an annealed condition.

Hopewellian metalworkers also cold worked, annealed, and ground pieces of copper into flat sheets of various thickness (.18 mm–1.25 mm inclusively) which were used alone or hammered together to produce finished artifacts that were mainly ornamental in form (Leader 1988; Wayman et al. 1992:102; Willoughby cited in Greber and Ruhl 1989:122). Sheet is thought to have been "cut" into desired shape through use of an embossing technique wherein bosses were impressed or grooved into one lateral surface of the sheet in the desired shape or design. The sheet was then turned over and the protruding bosses ground away, releasing the shape from the sheet. Shapes or designs could also be released by repeatedly abrading the inside of a grooved channel until it wore completely through (Leader 1988:102).

Sheets were used in various ways. Thick sheet (1.2 mm–2.9 mm by Willoughby's reckoning [Greber and Ruhl 1989:92]) was cut or formed into flat, undecorated, rectangular ornaments called simply "plates" or "breastplates" (Leader 1988:95–96; Willoughby cited in Greber and Ruhl 1989:91–92). Breastplates varied in size and in thickness. They often carried double perforations, presumably for attachment.

Thin, flat copper sheet was cut into various ornamental forms or "symbols." Artisans created geometric and/or zoomorphic designs. Their repertoires included antler, bear teeth, trefoils, swastikas, and intricate geometric shapes. Some zoomorphic sheets, which Leader (1988:99–101) calls effigy sheets, were cut into forms such as fish, birds' heads, and bear paws. Embossing and repoussé techniques were used to accentuate certain features of the forms, such as eyes or fins (Willoughby cited in Greber and Ruhl 1989). Some were perforated.

"Cutouts" are one of the most impressive and technically intricate ways in which thin copper sheet was manipulated by Hopewellian copper workers. Extremely intricate openwork designs were created on a sheet ornament using the removal techniques described above (Leader 1988:97; Willoughby cited in Greber and Ruhl 1989:106). Cutout designs featured sheared edges that were often rubbed, ground, and/or polished until the metal was very thin. Many of these objects were perforated, presumably for some type of attachment. They are found repaired with rolled strip copper rivets that held joining strips to the original ornament (Willoughby cited in Greber and Ruhl 1989:102, 112).

Other types of decorative ornaments were also fashioned from thin sheet. According to Willoughby (Greber and Ruhl 1989:104), single sheets were laid over a domed form (perhaps made of wood) and hammered into shape. In some cases, objects were also perforated. To produce hollow beads, sheet was bent around rounded forms (mandrels). Hollow bracelets were formed by removing strips from the sheet and hammering them along their long axis into rounded molds (called sinking), then hammering the protruding edges inward to form a c-shaped cross section (Leader 1988:86). They were then bent around a mandrel into final shape. X-radiography (Wayman et al. 1992:131) revealed that in one example, significant folding of the metal had taken place during the hammering phases of the manufacturing process.

Sheets were also used to laminate objects manufactured in other media. Several cruciform ornaments made of shell (and possibly reinforced with wood) were laminated with thin sheets of copper and of silver (Wayman et al. 1992:109). A copper headdress carries copper clad antlers made of wood (Willoughby cited in Greber and Ruhl 1989:99–100). Sheet seams are ground and burnished so as to appear imperceptible.

Bicymbal, or "yo-yo" shaped ear spool ornaments are probably the most complex copper artifacts manufactured in native northeastern North America. These highly visible, three-dimensional decorative artifacts are found in and around human remains in burial deposits as well as in other ritual contexts. They have aroused a great deal of stylistic and technological interest (Leader 1988; Ruhl 1992; Ruhl and Seeman 1998; Wayman et al. 1992; Willoughby 1903; Willoughby cited in Greber and Ruhl 1989). Ruhl's (1992:48) thoroughgoing study of these objects revealed that forms vary through time and over space based on the occurrence of a number of stylistic features, including construction technique. The manner in which ear spool discs were joined, particularly the use of various nonmetal adhesives to hold the components together, is among the many variations she identified. Using optical microscopy of several prepared ear spool cross sections, Wayman and colleagues (1992:101–103, 131) discovered that a number of the disks comprising the obverse and reverse of the finished spools were actually made up of at least three or four layers of copper sheet hammered tightly together. Importantly, the disks were riveted together at the midsection and joined at the outside edge through use of mechanical, rather than metal welding or bonding techniques.

Thus, in Hopewellian metalworking systems, although several types of solid objects are produced, manufacture and manipulation of sheet stock become an important aspect of artifact production. Repeated annealing and hammering sequences, as well as considerable folding, perforating, grinding, and polishing, were important techniques used in artifact fabrication. Embossing, cladding, and sinking techniques were added to the technological repertoire of earlier times. First evidenced in Archaic metalworking traditions, the use of rolled rivets for joining and repair becomes more frequent. Twine wrapping and adhesives are now also used as joins (Ruhl and Seeman 1998:659). Wayman and colleagues (1992:112–113, 133) found no evidence of casting in this industry, although they postulate that some objects are very likely to have melted in funerary fires after manufacture. Several types of finished artifacts appear to exhibit standardization and development in form and design during this period, although Ruhl (1992:56) points out that there remains variability in the dimensions and shapes of artifacts similar in design. The question of the use of templates to achieve that standardization has not been demonstrated conclusively (Ruhl and Seeman 1998; Greber and Ruhl 1989:141). Also, the purposeful selection of high arsenic copper (Wayman et al. 1992:130) and the intentional use of copper and silver cladding techniques point to the importance of color and visual effect of the finished product (see Ruhl and Seeman 1998:655, 657, 659). Overall, the impressive quality and complexity of the metalworking found among Ohio Hopewellian peoples have prompted Leader (1988:198) and John Halsey (1996:15) to suggest that metalworking specialists worked at manufacturing copper items, possibly on a full-time basis (Leader 1988:198).

Less is known with certainty about where the raw copper actually came from, where objects were fashioned, or how distribution of objects and/or raw material took place. Stuart Struever and Gail Houart (1972:66–67) have hypothesized that the Trempeauleau site in the Upper Mississippi Valley might well be pivotal in the interaction sphere. They postulated that this was due to the heavy presence of finished objects and manufacturing debris there as well as its proximity to the Lake Superior sources. Based on the amount of copper at the Hopewell site (Scioto River, Ohio), they also posited that the Scioto region was a "concentrating production area for this material" (1972:66). Seeman's (1979:385) later investigations demonstrated that both materials and finished objects were probably moving in and out of both the Havana and Scioto regions. However, distributions of particular objects within and across regions are far from uniform, with scrap from manufacturing found in both areas (Winters 1981:20). Dan Morse and Phyllis Morse (1983:163, 166, 167, 171) have noted that ritual copper artifacts found in the central Mississippi Valley strongly resemble Illinois Hopewell, and may have been carried to sites there from regional centers in Illinois or even farther east.

Importantly, in her study of copper distribution and interregional exchange during this period, Goad (1979, 1980) has discovered that not all of the copper artifacts from the southeastern Hopewellian complexes she analyzed were made from Lake Superior copper, as had long been assumed (see Levine 1996). The results of her optical emission spectroscopic provenance studies revealed that some were fabricated from more local resources in southeastern Tennessee, northwestern and southwestern North Carolina, and northeastern Georgia. She (1979:245) suggests that copper from the Great Lakes may have entered the region via the Copena regional centers located in the Tennessee River Valley region of northern Alabama. As time passed, Lake Superior copper importation waned as exploitation of local resources and adoption of copper working technology escalated among southeastern Hopewellian complexes.

Mississippian Copper Working (A.D. 900 to European Contact in Some Areas of the Southeast)

While the Late Woodland, Fort Ancient, and Oneota cultures of the midcontinent are not noted for extensive copper use (Halsey 1996:15), the Mississippian represents refinement and sophistication in craftsmanship. Although recovered in greatest quantities at Spiro, other major Mississippian centers such as Etowah and Moundville have also yielded a considerable variety of copper items (Brown 1985; Sampson and Esarey 1993:452). Copper artifacts are also found in much smaller quantities at various other sites throughout the Mississippian sphere. Some of the copper for Mississippian artifacts is thought to have been procured by Mississippian peoples from Great Lakes sources; however, a shift is noted to primarily southeastern resources (Goad 1980:271).

Copper was exploited by Mississippian peoples for internal and external consumption and was traded over long distances, with Cahokia being one of the trade centers during its heyday (Brown, Kerber, and Winters 1990:271; Leader 1988:185–187). Sufficient material evidence exists for the presence of a workshop at Etowah where Leader (1988:173–174, 199) suggests that specialists were at work, possibly full time, making and repairing copper items.

Throughout the Mississippian, copper is used for fashioning ceremonial artifacts and implements that are thought to be markers of high social status, authority, and wealth (Brown 1985; Leader 1988:198). James Brown, Richard Kerber, and Howard Winters (1990:272) claim that along with marine shell artifacts and pearl beads, copper is now at "the apogee of value" as a prestige mortuary good. It may have been somehow controlled by the elite (Leader 1988:199). Adornment items from the period include hair ornaments, gorgets, beads, headdresses, ear spools, and copper-covered masks. Ritual items are found in the form of zoomorphic and anthropomorphic rattles and effigies. Spear heads and other ceremonial implements such as celts, large axes, bipointed needles, and pins are also found in the industries (Hamilton, Hamilton, and Chapman 1974). Many of these objects, as well as the dramatic, intricately executed repoussé copper headdress plates so well known from elite mortuary contexts of the period, are important material symbols of the Southeastern Ceremonial Complex (SECC), a religious, symbolic, and exchange system that stretched over broad areas of the Southeast and Midwest during the Mississippian and early contact periods (Brown 1985:117; 1989; Muller 1989:11, 25). As Kelvin Sampson and Duane Esarey (1993:452) have argued, copper was clearly one of "a few central media used for display of the paramount Mississippian period symbols."

Mississippian metalworkers still fabricated solid implements and sheets in the manner of earlier metalworkers. Refinements center on the further thinning of sheet into foil (defined by Leader [1988] as sheet thinner than .05 cm), and the extensive use of repoussé technique in executing design motifs. The cutout technique is also documented. Covering objects made of other materials such as wood and stone with thin copper sheathing not only continues, but also intensifies during this period, becoming a significant feature of Mississippian metalworking. Human and animal effigies, rattles, ear spools, bodkins, finials, beads, bladelike objects, and sun gorgets are examples of clad objects (Leader 1988:129–135, 198; Hamilton et al. 1974:176–177).

Copper sheets and foils were also used for the manufacture of elaborate ceremonial, symbolic, and status-related ornamental artifacts such as highly decorated plates and badges. Sheet artifacts were now larger due to use of delicate riveting to join multiple hammered sheets. Plates were elaborately embossed with iconographic motifs that reflected important themes associated with the cult complexes they represented (Brown 1985:114). These plates also varied in artistic style across the Southeast according to particular design principles (Brown 1985:98–99). Designs became more standardized through use of templates; like plate styles are found widely distributed over long distances across the Southeast (Brown 1989). Mixed styles at individual sites indicate trade in these prestige objects (Brown et al. 1990:265).

Aside from Leader's (1988) careful examination of materials from the Etowah site and from Mississippian collections in Florida, it appears that little archaeometric investigation has been carried out on Mississippian materials. He has noted continued, but refined use of already-familiar techniques, including embossing, repoussé, riveting, and cladding; he also documents perforation, shearing, grinding, molding, and burnishing. These techniques were augmented by more complex methods of attachment and repair. Some of the artifacts he examined, including headdresses found in mortuary contexts, are thought to have been used and even patched with pieces of other plates before placement in graves (Brown et al. 1990:265; Hamilton et al. 1974:187; Leader 1988:184). Henry Hamilton and colleagues (1974:173) have suggested that this type of intensive refurbishment may indicate that insuring the continued existence of particular plates might have been of utmost importance to Mississippian peoples.

Schroeder and Ruhl (1968:165–166) have conducted a small-scale metallographic investigation into the manner in which Mississippian copper sheet was manufactured. Responding to claims by previous researchers that sheets were manufactured without annealing, their inquiry demonstrated that the metal in one sheet specimen had indeed been subjected to a high temperature (about 800° C) anneal. The other sample, taken from a plaque fragment, indicated that some cold working had been performed as a final shaping step on a previously annealed sheet.

The Native Copper/European Copper-Base Metal Transition

In the Northeast, native copper working extended into the contact era. Recently, archaeologists have confirmed archaeometrically that in the early stages of native/European interaction, at least in some locales, artifacts manufactured of native copper did indeed occur alongside artifacts of like form fashioned from European-derived copper-base metals (Abel, Burke, and Stothers

2000; Dunbar and Ruhl 1974; Fox, Hancock, and Pavlish 1995:282). However, European-introduced copper-base metals in the form of smelted copper and brass, an alloy of copper new to native peoples in color and composition, rapidly came to dominate native copper-base metalworking systems. In many instances, European metals and native-fashioned artifacts began to appear in material repertoires of native peoples not known for their metalworking traditions (Bradley and Childs 1991:7).

THE INTRODUCTION OF EUROPEAN COPPER-BASE METAL INTO INDIGENOUS SYSTEMS

The European/Amerindian Copper-Base Metal Trade

Smelted copper-base metal appeared in the Northeast in the late second half of the sixteenth century, primarily in the form of copper and brass kettles. A major object of exchange, kettles rapidly became one of the most important European commodities traded to Native Americans. Early, iron-banded "red copper" kettles were technologically elaborate and visually distinctive. Appearing in the late sixteenth century, these heavy, diagnostic "Basque" type kettles were exchanged with native peoples of eastern Canada for furs, hides, and skins (Fitzgerald 1995; Fitzgerald, Turgeon, Whitehead, and Bradley 1993; Turgeon 1997). Between 1584 and 1587 alone, over 500 were traded in to the St. Lawrence (Fox 1991:3). After the turn of the seventeenth century, other less complex, undoubtedly more economical, rolled-rim styles of copper kettles replaced this early type. Basins, bracelets, bells, ear rings, awls, needles, beads, and rings rounded out the early native/European exchange repertoire of copper-base metal trade goods (Turgeon 1997; Fitzgerald et al. 1993).

Along with copper kettles, two types of brass or "yellow copper" kettles were manufactured in Flanders and in northern France for trade. Simpler in design than their red copper counterparts, archaeologists have postulated that brass kettles appear to have replaced the more elaborate copper kettles in the Canadian trade soon after the turn of the seventeenth century (Fitzgerald et al. 1993; Fitzgerald and Ramsden 1988). In contrast, Norman brass was the first trade kettle metal to appear on Iroquoian and Susquehannock sites (Bradley and Childs 1991). Brass bells and buttons were also traded (Cleland 1971).

Tracing kettles to their specific European production centers has been difficult (van Dongen 1995:127). Copper for the manufacture of exported kettles came largely from Scandinavian and Germanic countries and was worked by

Flemish and French coppersmiths (Turgeon 1997). In the seventeenth century, higher quality, cheaper copper than could be produced in Europe was also shipped to Amsterdam from Japan (Perrin 1979:9). Amsterdam merchants also handled copper worked in Aachen, western Germany, and in Dutch cities (van Dongen 1995:116). Brass was produced with zinc mined locally in Flanders and copper imported from central Europe.

Surprisingly, no inclusive formal or technological typology currently exists for imported kettles. Two basic kettle forms appear in contact period northeastern North America, hemispherical kettles (commonly known as "trade" kettles), and the much less frequently found two-part cylindrical kettles (van Dongen 1995:116–117). Hemispherical kettles were achieved largely by one of two basic fabrication processes, raising (by hammering and planishing) or spinning (Bradley 1987:197–199; Untracht 1968). Both processes involved working a single cast sheet into a basin or flattened bottom basin shape. The edges of the basin were worked outward and fitted around an iron support ring. Most were plain and unmarked. Bradley (1987:197–199) identifies a French-related variant of this type that is not "standard" in size and has distinctive type lugs. These forms exhibit more patterned battery work and makers' marks.

The other type, two- or multiple-part cylindrical kettles, were called Dutch "domestic" kettles (Bradley 1987:197–199). Generally made of thicker gauged sheet (> 1.0 mm), these kettles had straight sides made up of one or more bent sheets, and a flat bottom. Parts were held together by nails or rivets and riveted bands. Seams were caulked with lead or tin.

Both copper and brass kettles were outfitted with iron bails (handles) that were attached to the kettle body through the use of lugs riveted to the kettle. Lugs themselves exhibited variability in form and manner of attachment. Often manufactured in graduated sizes, finished kettles were shipped to New France from LaRochelle and Bordeaux (Turgeon 1997). They were nested for economy of transport both on ships and in New World trading canoes.

Native Appropriation and Use of Copper-Base Trade Metal Artifacts and Materials

What has become clear to researchers is that native peoples often consumed copper-base metal trade artifacts for very different material, metaphorical, and symbolic purposes than Europeans intended at the time of their original manufacture (Fitzgerald 1995:36; Martin 1975). At first, even the most elaborate, expensively manufactured, and distinctive European copper-base kettles were not sought after by native peoples simply to replace purportedly "inferior" native-manufactured clay pots used for centuries for food preparation. Rather, when used at all for this purpose, kettles were reserved for special communal feasts (Turgeon 1997).

Copper-base metal kettles also played important material and symbolic roles in varied alternative contexts. According to Alexandra van Dongen (1995:142) the earliest attraction to European kettles was for the metal they contained. As many authors have noted, kettles were cut up and the metal reworked to make ornaments or utilitarian objects. Somewhat later, unaltered kettles were more frequently found in mortuary contexts rather than in domestic ones, serving as containers for human bones or funerary offerings like beads or foods (van Dongen 1995:143; Kenyon 1982; Martin 1975; Turgeon 1997). Huron, Petun, and Neutral reburial Feasts of the Dead also featured both broken and complete kettles and kettle parts, as well as other copper-base metal artifacts such as coils of hollow tubing, rings, beads, clips, and rattles, as important grave offerings (Kenyon 1982; Martin 1975; Turgeon 1997). Kettles also served as drums during the ceremony, part of which was called "the kettle" (van Dongen 1995:157).

In some cases, kettles placed in grave contexts are found to have been purposefully destroyed (Martin 1975). They are thought to have been ritually "killed" as a part of the funeral ritual, therein releasing the soul of the deceased while providing him/her with the "soul" of the kettle for use in the afterlife (van Dongen 1995:156; Martin 1975:115–116). In others, kettles were deposited intact in graves, also so that their "spirits" could accompany the deceased to the "other world" (van Dongen 1995:156).

In one mid-seventeenth-century encounter Acadian merchant Nicolas Denys reports to have had with a Micmac man, the man attests to native belief that kettles were indeed animate and of use to individuals in the other world. The exchange involved the opening of a native grave in which a corroded kettle was found. It was struck to test its resonance. Upon hearing that it "no longer sounded," the man explained that

> they [the deceased] have need of it, since it is among us a utensil of new introduction, and with which the other world cannot [yet] be furnished. Do you not indeed see, that it no longer says a word, because its spirit has abandoned it to go and be of use in the other world to the dead man to whom we have given it? [Denys 1908 cited in Martin 1975:115; see also van Dongen 1995:156].

Most importantly for this work, native peoples acquired kettles and other types of copper-base trade items and reworked them into new objects that they fitted into their own material, technological, and ideological systems. Before ever having been "used up" from or even used for cooking, trade kettles were frequently broken down and the sheet metal cut up into smaller pieces for further exchange or gift giving. It was also reformed into items of bodily adornment such as rolled beads, conical tinkling cones, bracelets, and rings. Finished items are also likely to have been traded (Drooker 1996b). In some cases, reworked finished objects are thought to have replaced, in form and/or in function, native antecedents in materials such as bone, shell, or stone. In others, new objects were created that had no formal and/or functional antecedents in native material repertoires. Newly fashioned objects were worn on the body or attached to clothing in battle or in diplomatic, funerary, and other ceremonial or festive contexts. While the purpose of most of the early reworking appears overwhelmingly to have been to produce nonutilitarian items, objects such as saws, projectile points, needles, awls, and cutting and scraping tools were also fashioned (Anselmi, Latta, and Hancock 1997; Bradley 1987; Cleland 1971; Turgeon 1997).

Whether and to what extent the appearance of imported copper-base metal transformed native metalworking practices is now being explored systematically. Archaeologists have described Huron techniques of initial kettle reduction and basic reworking of the metal (Anselmi 1994; Latta, Thibaudeau, and Anselmi 1998). These techniques include cutting, folding, rolling, cold hammering, annealing, and edge finishing by grinding. Use of "scrap" or purportedly discarded "wastage" as expedient tools has also been documented through identification of use wear (Latta et al. 1998). Metallographic analysis has provided evidence for primary cold working, annealing, and bending (rolling). This evidence comes from archaeometallurgical study of two specific artifact forms, spirals and hoops, which are narrowly distributed spatially and temporally in the Northeast (Bradley and Childs 1991; Childs 1994).

Archaeologists have also used elemental analyses to identify reliably the composition of copper-base metal materials on protohistoric sites and to address questions of copper-base metal material adoption, distribution, and use. As explained in chapter 3, trace element analyses carried out through use of instrumental neutron activation analysis (INAA) have provided reliable evidence for distinguishing native copper from European copper-base metals. In addition, "lots" or "batches" of similar kettle chemistries have been identified for the purpose of tracing intrasite and intersite distribution and movement of materials within and across regions (Hancock, Pavlish, Fox, and Latta 1995). Research attempts to seriate temporally copper-base kettle/sheet metal chemistries for use as chronological markers in dating protohistoric sites also hold promise (Hancock et al. 1994; Walker, Hancock, Aufreiter, Latta, and Garrad 1999).

Symbolic Underpinnings for Copper-Base Metal Appropriation in the Protohistoric/Early Contact Period

Native peoples' attraction to European copper-base metals and metal objects during this period had functional, aesthetic, and ideational dimensions (Fox 1991; Martin 1975; Turgeon 1997). It is thought to have been particularly attractive to Native Americans not only for its utilitarian worth but also for its brilliant red color, its excellent working properties, its symbolic associations, and its importance in native ritual systems (Hamell 1983, 1986; Jaenen 1974; Martin 1999; Turgeon 1997). Of early Micmac practical attraction to European trade kettles, Nicolas Denys (Jaenen 1974:276–277) reports,

> They have abandoned all their own utensils, whether because of the trouble they had as well to make and use them, or because of the facility of obtaining from us, in exchange for skins which cost them almost nothing, the things which seemed to them invaluable, not so much for their novelty as for the convenience derived therefrom. Above everything the kettle has always seemed to them, and seems still, the most valuable article they can obtain from us.

Ethnohistorian Cornelius Jaenen's (1974) critical examination of early seventeenth-century native attitudes toward French culture and French goods also reveals that from a utilitarian perspective, native people were impressed by and "greatly" appreciated items of European technology, especially blankets and copper kettles. According to Jaenen (1974:275), "all tribes showed an appreciation of the knives, hatchets, kettles, beads, cloth, and eventually, the firearms of the French." At the same time, however, he (1974:276) is careful to point out that "there is reason to believe that the Amerindians valued European trade goods such as beads, mirrors, bells, and caps, for their aesthetic, magical, or purely decorative and fascinating worth, not their economic value."

As they had done for native copper, it is clear that native people also recognized European copper-base metal for its color, its purity, and its excellent working properties. The copper-base metal material itself was easily worked and had excellent reflective qualities. Naturally corrosion resistant, its luster could be brought back simply by rubbing (Kellogg 1968:345; Turgeon 1997).

Significantly, very early hints as to the "worth" of copper-base metals to native peoples, its use as ornaments, and hence a native motivations for trade for it are provided by Giovanni da Verrazano (see Levine 1996:47). As early as 1524, Verrazano describes the appearance and preferences of native peoples of Narragansett Bay in the following manner:

> We saw upon them several pieces of wrought copper, which is more esteemed by them than gold, as this is not valued on account of its color, but is considered by them as the most ordinary of the metals—yellow being the color especially disliked by them: azure and red are those in highest estimation with them.
>
> Of the things which we gave them, they prized most highly the bells, azure crystals, and other toys to hang in their ears and about their necks; they do not value or care to have silk or gold stuffs, or other kinds of cloth, nor implements of steel or iron [Winship 1905:15–16 cited in Hamell 1986:80].[4]

In the late 1660s, Father Claude Allouez (Thwaites 1896–1901:50:265) would report from the shores of Lake Superior:

> The Savages revere this Lake as a Divinity, and offer it sacrifices. . . . One finds at the bottom of the water pieces of pure copper, of ten and twenty livres' weight. I have several times seen such pieces in the Savage's hands; and, since they are superstitious, they keep them as so many divinities, or as presents which the gods dwelling beneath the water have given them, and on which their welfare is to depend. For this reason they preserve these pieces of copper, wrapped up, among their most precious possessions. Some have kept them for more than fifty years; others have had them in their families from time immemorial, and cherish them as household gods.

The cognitive and ideological dimensions underpinning the attraction for and perceived value of particular trade goods to native consumers have been pointed out by Wilcomb Washburn (1967). These concepts have been elaborated upon most eloquently in works by Christopher Miller and George

Hamell (1986) and Hamell (1983, 1986). Hamell (1983:5) has argued that for native Siouan, Algonquian, and Iroquoian groups of the Northeastern Woodlands, some trade materials, copper among them, were "symbolically charged substances" that expressed important metaphysical concepts of "light" and "life." As one of several reflective materials, including shells and crystal, copper was integrated into the ideational worlds of indigenous recipients according to native systems of internal logic, meaning, and value (Hamell 1986:73). Hamell (1986) constructs what he thinks constitutes this logic. He claims that in the native ideational world, certain colors, particularly white, sky blue, green, red, and black carry particular ideational and aesthetic values and functions in the tripartite axial division of the native cognitive world (Hamell 1986:324–325). These values are represented in materials like shell, catlinite, native copper, and in natural pigments such as blue green clay, red ochre, and charcoal. Particular colors, especially white and red, and their material manifestations reflect important life-related wealth concepts such as physical, social, and spiritual well being. Specifically, red mediated between light (life) and dark (death). Suggestive of blood, red color meant high emotion, excitement, and power (Hamell 1986:325).[5] Since these states of being were considered positive and animate, accumulating them meant gaining wealth as a kind of medicine that carried with it life-affirming and restorative qualities, such as long life and success in important activities such as hunting, courtship, and warfare. As such, they were important ingredients of medicine bags (Fox 1991:5; Hamell 1986:76–77). This kind of wealth could be used or displayed as ornamentation or as gifts to the living or in grave offerings. It had exchange value as a prestige, luxury, and status good.

These categories of wealth are ascribed cosmologically to the "other world" —specifically to the underwater world, from whose keepers humans receive such wealth in reciprocal ritual exchange (Hamell 1986:75, 79). According to Allouez (Thwaites 1896–1901:50:289), "They [native peoples] say also that the little nuggets of copper which they find at the bottom of the water in the Lake, or in the Rivers emptying into it, are the riches of the gods who dwell in the depths of the earth." Horned or antlered underworld manitou or "grandfathers" dwelt in such places as the water's or forest edges considered metaphorical thresholds to the "other world." In the ethnographic record of some Western Great Lakes Algonquians, one specific spirit, Mishipizheu (also Mishi Bizi, or Missibizi) is associated with the (under)waters, fishes, stormy waters, and terrestrial game (Fox 1991:5; Martin 1999:200–204). Sometimes conflated with Mishi Bizi is another related underwater creature, the fearsome and powerful Great Serpent, or Horned Snake. This creature had copper scales, copper antlers, and a powerful, destructive copper tail. Native peoples cut red copper from its horns when it rose out of the water (Kellogg 1917:105; Fox 1991:5). These manitou were the sources of gifts, charms, and red copper, and were in charge of dispensing it. Gifts of copper had to be appropriately acknowledged through ceremonies and placement of offerings near Lake Superior or in burial contexts. Offerings could also calm the waters, avert disease, or insure good fishing or safe passage on the lake (Martin 1999:202).

Hence, for Miller and Hamell (1986), when items made of materials such as copper or glass were introduced to native peoples, they were not seeing something "new"; but rather they were encountering materials aesthetically similar and also "other worldly." For these reasons, these items could be and were assimilated easily into existing material inventories and ideological systems right alongside native copper, siliceous stones (crystals), and shells for use in familiar social, ceremonial, and ideological contexts (Hamell 1983:18; Miller and Hamell 1986:315, 318, 326).

Bradley and Childs (1991:16) hypothesize such material and symbolic congruence in their technological analysis of trade copper-base metal spirals and hoops found among the late sixteenth-/early seventeenth-century Susquehannock and Five Nations Iroquois (discussed in chapter 2). They propose that there may well have been cognitive associations between native belief in the restorative qualities of the copper from the Underwater Panther's tail (Mishi Bizi) and adoption and use of European copper for the manufacture of spirals and hoops. Bradley and Childs reason that if pieces of the Panther's tail (whether European or native copper) are considered valuable charms possessing significant healing powers, having wealth in this form would be especially sought after in early contact times, when epidemic disease was so prevalent and destructive and native peoples were so powerless to curtail its spread (see also Penney 1985).

Interestingly, however, native attraction to European-derived copper-base metal appears to have declined through the early contact period. For archaeologists, this is marked by decreased appearance of copper-base metal items in burials and its increased occurrence in nonceremonial or nonsymbolic contexts (Drooker 1996b:172). Perhaps this is due to the increased availability of the material to the point that native markets became saturated and its economic and social value as a rare and exotic good declined. What impact this had on how native peoples perceived copper as a "symbolically charged substance" is unexplained.

The Search for Meanings of Copper in Prehistory

Hamell (1986:79) argues that the particular "mythical reality" of copper-base trade material acceptance he proffers for the contact period represents a pan-northeastern native ideology that long predates European contact. It is, however, difficult to test the historic depth of his claims, especially as we reach deeper into prehistory. This is not to dismiss them, however, since a number of themes, especially those relating to the importance of the color red, and the associations of copper with underworld grandfathers and underwater creatures may indeed have material representations in the prehistoric record that bear further investigation. David Penney (1985:185, 188) has discussed some important cosmological connections among the motifs and ceremonial regalia of prehistory such as underwater creatures (fish, horned serpents) and antlered or horned headdresses. While these and other culturally significant motifs were frequently executed in copper, establishing with confidence clear symbolic linkages among copper, symbolic motif, and prehistoric world view is beyond the scope of this work.

However, from a technological point of view, prehistoric peoples clearly did recognize early on and exploit copper's exceptional working qualities and its utility as a material for artifact production and visual display. Archaeometric analyses have demonstrated that the metal was worked largely through cold and hot working and annealing, and a variety of what might be classified as "secondary" tooling and forming techniques (after Franklin et al. 1981).[6] This technological knowledge appears to have been widespread, having transcended social, geographical, and temporal boundaries. In a metallurgical sense, these processes never "evolved" to true full-blown "metallurgy" through which metals are "produced" and formed through pyrotechnical means. Some techniques, such as working delicately with sheet, connecting it in careful ways to make large, flat objects, were refined through time as witnessed by the elaborately executed Mississippian foil plates. Some, such as assembling multiple sheets in complex and unique ways to create Hopewellian ear spools, were not seen again in prehistoric or protohistoric metalworking repertoires.

Importantly, archaeologists have long agreed that from its earliest use in the Archaic period, copper was indeed considered a valuable material. Winters (1968:181), for one, has argued that copper was valued by prehistoric peoples as a raw material, not for the artifacts produced from it. His arguments (Winters 1968:204), as those of Struever and Houart (1972), are based on the notion that for many peoples, copper is exotic and must be imported; as a rare commodity, it is considered an important form of wealth that is differentially distributed.[7] What particular ideational meanings may have underpinned its desirability or its special status as a raw material are not considered.

Copper's value in native material and symbolic systems has been demonstrated repeatedly through analyses of its use contexts, its distribution, and its sourcing. As a major repository for valued, exotic materials, copper artifacts as ornaments or tools, either used or unused, have appeared in burial contexts for millennia. Their placement with the "favored dead" in elaborate funerary displays attests to its special value and its meaning as a status marker (Penney 1985:167; Winters 1968). For instance, the association in Mississippian burial contexts of very finely executed iconographic copper plates and badges with important personages carried important cosmological messages concerning the distinctive, elite, even sacred status of the wearer in life and in death (Brown 1985; Penney 1985). Even the "expressive" qualities of presumably utilitarian implements made of copper are conveyed vividly by their contextual deposition (Childs 1994; Wells 1991:90).

LOOKING AT COPPER WORKING PRACTICE OVER THE *LONGUE DURÉE*

Throughout prehistory, copper was worked by metalworking craftspeople through cold hammering and annealing processes with some hot forging. Ursula Franklin and colleagues (1981:38) have called this phenomenon a "single technological pattern . . . [that] persisted through time and cross-cut traditionally recognized cultural and linguistic boundaries." They have termed this the *primary* technique of native copper working. Yet, within this ubiquitous *primary* copper working method, particular copper working traditions developed using different, distinctive *technological styles* (in Lechtman's [1977] sense of the term). The change from hammering *nodules* of copper into solid, heavy finished objects to the manufacture of *sheets* for use as stock from which to fabricate artifacts is one example of a major shift made early on in native copper working technology. This shift persisted through time. Refinement of sheet production and the elaboration and standardization of design and execution of artifacts made from it are another. Continued use of introduced metal in sheet form but with a very different final product in mind represents both continuity and cleavage in both conceptual processes and production technique.

It appears, however, that variation in copper working *styles* had more to do with differences in the use of what Franklin and colleagues (1981) have called *secondary* working or *tooling* techniques. These would include such practices as riveting, perforating, shearing, grinding, joining, and sinking. Individual copper working traditions developed, implemented, and refined particular combinations of these techniques that gave products a distinctive stylistic character. Some of these secondary techniques were unique and quite innovative. Some did not persist over time. Why particular techniques emerged, why and how they might have varied, and why they were chosen had to do with many factors, including in what form the material was actually procured, the skills and knowledge of the artisans, the mechanical limitations of the material itself, and the uses for which the artifact was intended. These all reflect the variable values and meanings native peoples placed on the material and on the object. In turn, these values and meanings take and maintain form within particular, material, social, economic, ideological, and historical settings. What is clear, however, is that through prehistory and the early contact period, copper maintained important symbolic and material value to the peoples who used it. Only in the historic period did things begin to change. Throughout the greater history of its use, however, copper often appeared far from its sources in forms and contexts that had highly symbolic and communicative meanings. While these meanings are not always clear, as Childs has said (1994:232), the belief system of copper using and copper circulating peoples has undoubtedly "strongly influenced" the relationships among native peoples, copper, and its uses.

5 / "Lost Sheep . . . in the Jaws of the Wolf"

The Mid-Seventeenth-Century Illinois in Ethnohistorical and Archaeological Perspective

INTRODUCTION

In the early documentary record of European contact in the Western Great Lakes/Upper Mississippi Valley, the Illinois emerge as one of the most influential and historically visible native groups to occupy the region's cultural landscape. As their encounters with French traders and missionaries intensified throughout the mid- to late seventeenth century,[1] they are seen to have become increasingly forceful political and commercial players both within and outside their own territories. Actively seeking interaction with the French, the Illinois would eventually establish themselves as "middlemen"[2] in native trade into the interior. They would also remain loyal and valuable allies of the French in military actions against mutual enemies.

For archaeologists, isolating the Illinois in the protohistoric archaeological record in order to illuminate these dynamics has been a rather frustrating enterprise (Emerson and Brown 1992; Walthall and Brown 2001:88). However, recent archaeological investigations at the Haas/Hagerman site (also known as the Iliniwek Village historic site), Clark County Missouri, have finally brought the historical record and the archaeological record into line. The mid-seventeenth-century Illinois have now been linked conclusively to this site, where the archaeological assemblage has yielded traces of early European material influence—iron, copper-base metal, and glass. This evidence shows clearly that during the time of its occupation (1640–1682) the Illinois were obtaining particular types of European-introduced materials, particularly copper-base metals, and integrating them into their cultural milieu. Weaving together pertinent documentary evidence from roughly contemporary source materials and archaeological evidence from Haas/Hagerman and other sites creates a highly textured material, spatial, economic, historic, and sociopolitical fabric within which these activities took place (see Figure 5.1 for sites and locales mentioned in the text). It places the copper-base metal sample used in this study not only in its depositional, distributional, and use contexts, but also in the larger picture of initial encounter, exchange, and early material and social change among the Illinois.

THE PROTOHISTORIC ILLINOIS IN THE WESTERN GREAT LAKES AND ILLINOIS COUNTRY

Cultural Identity and Origins

Historically, the Illinois are one of several Central Algonquian groups who occupied what was to become known as "the Illinois Country" and the regions west of Lake Michigan in the mid- to late seventeenth century. They are described by Charles Callender (1978:673) as a "group of independent tribes united by a common language and sharing a tradition of common origin."[3] He characterizes Illinois culture as "transitional," combining Prairie Siouan and Central Algonquian elements, even revealing some traits traceable to the Southeast. Ethnohistorian Raymond Hauser (1973:7) refers to the Illinois as "Prairie Algonkian," while Emily Blasingham (1956:218) posits that their lifeways appeared more closely aligned with those of prairie groups than those of Ohio Valley peoples.

Some uncertainty surrounds just how many subtribes, or tribal "villages" comprised the Illinois in the mid- to late seventeenth century (Bauxar 1978:594). Until direct French contact was made with them in their own territories in 1673, most contemporary accounts simply refer to the Illinois and their numbers in general terms. In the early 1680s, French explorer René-Robert Cavelier, Sieur de La Salle (Margry 1879–1888:96, 201) provides two similar reports on the composition of the Illinois confederacy. At one point, he mentions 10 *familles* of Illinois, including the Peoria, Kaskaskia, Tamaroa, Coiracoentanon, Chinko, Cahokia, Chepoussa, Amenakoa, Oouka, and Acansa, and at another, he mentions 10 "nations," repeating the first seven and this time adding the Moingwena, Tapuaro, and Maroa. Deliette (1687–1706) mentions only seven, the Peoria, Kaskaskia, Tamaroa, Coiracoentanon, Tapouara, Moingwena, and Cahokia (Pease and Werner 1934:341–342). Among modern analysts,

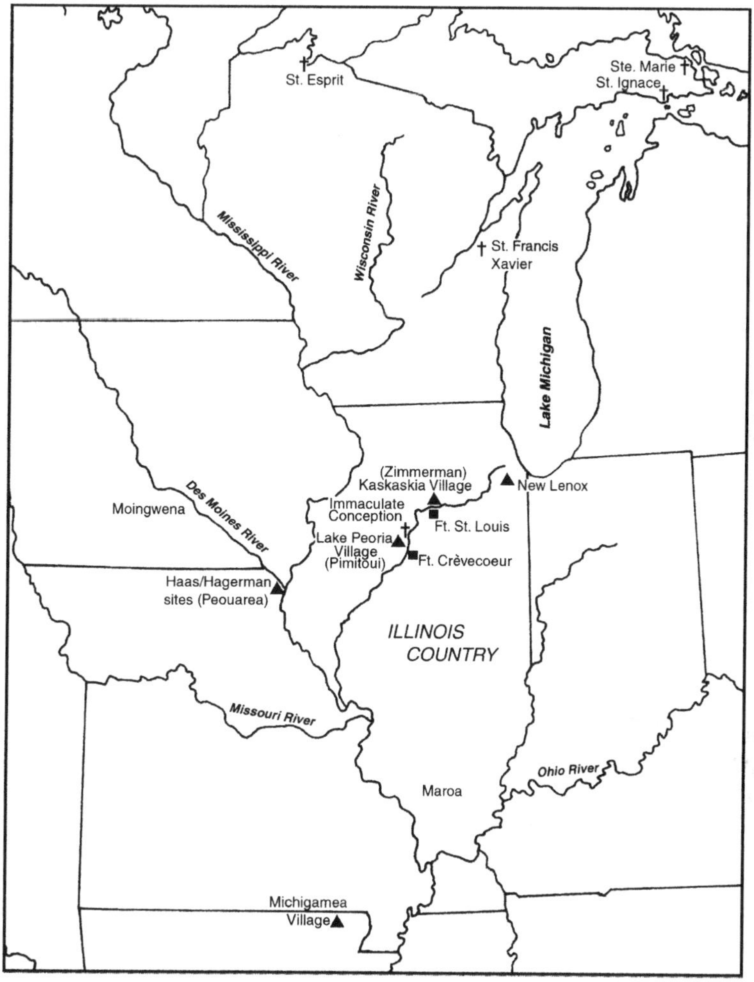

Figure 5.1 Illinois villages, French outposts, and other locales in the mid-seventeenth-century midcontinent, mentioned in chapter 5.

Margaret Brown (1979:227) lists as many as 17 "villages," and Joseph Bauxar (1978:594) lists 12. Joseph Zitomersky (1994:102–103) mentions three additional problematic designations, the Mosopelea, the Mishibousa, and the Medchipouria, which he concludes refer to the same group that may have been part of the Illinois at the close of the century. He (1994:107) also points out discrepancies in the English translation of the original French "maroa" to "tamaroa," further complicating an already hazy historical picture.

Algonquian speakers, the Illinois are closely related linguistically and culturally to the Miami, another Central Algonquian group of related tribes that inhabited the area around the southern tip of Lake Michigan in the mid-seventeenth century (Callender 1978:673–674; Kinietz 1996:161–162). Both groups spoke very closely related, mutually intelligible dialects of the Miami-Illinois language (Callender 1978:673; Costa 2002:2, 2003; Pease and Werner 1934:307). Hauser (1973:6, 21, 128) has emphasized the commonalities between the two groups, claiming that during this period they were indistinguishable linguistically and culturally. He suggests that they may have been united at one time, having separated shortly before contact (see also Callender 1978:673). Indeed Vernon Kinietz (1996:162) uses data from what is known of the Illinois to fill gaps in his description of early historic Miami culture. Callender (1978:674) is careful to point out, however, that despite their cultural and linguistic affinities, relations between the two groups were often hostile (see also White 1991:29, 34).

The original homelands of the Illinois are unknown. However, archaeological and ethnohistorical evidence argues strongly against in situ development in the Western Great Lakes and Upper Mississippi Valley. When and under what circumstances the Illinois entered the regions west of Lake Michigan and the lands surrounding the Mississippi River to the south and west is uncertain. It appears increasingly likely, however, that they migrated westward sometime in the early to mid-seventeenth century from regions somewhere farther to the east, most probably from the areas surrounding Lake Erie (Callender 1978:678; Thwaites 1896–1901:55:101). While archaeologists continue to explore affinities in ceramic styles between this region and the Illinois Country, they have not been able to trace specific migration routes.

Illinois Settlement Distribution in the Western Great Lakes

Earliest reports of the Illinois in the Western Great Lakes and in the traditional Illinois homelands to the south are vague, fragmentary, and indirect. In 1640, the Eriniouai (Illinois) are reported to have inhabited a vast area beyond the St. Lawrence and Lake Huron along the banks of Lake Michigan in what now encompasses northern Illinois and Wisconsin. They shared the area with the Winnebago, Sioux, Assiniboine, Mascouten, and Potawatomi (Jablow 1974:I:1; Jaenen 1996:118; Zitomersky 1994:77). According to Zitomersky (1994:78) they may have been located specifically on lands surrounding southwestern Lake Michigan. In the mid-1650s, Jesuit missionary Father Gabriel Dreuillettes records that Illinois population in the region west of Lake Michigan was 100,000 individuals from 60 villages (Thwaites 1896–1901:41:247). However, some scholars now argue that these numbers are inflated, maintaining that they may reflect more groups than the Illinois alone or represent a winter dispersal settlement pattern rather than an enumeration of more nucleated summer village composition (Jablow 1974:I:3; Temple 1966:13; Zitomersky 1994:78). As both Wayne C. Temple (1966:14) and Joseph Jablow (1974: I:15) have noted, population estimates and settlement descriptions recorded by the missionaries at this early time are likely to have been inconsistent since they had never traveled to the Illinois Country themselves. Despite disagreements as to their actual numbers, the Illinois are thought to have been the most populous "nation" occupying the region at this time (Jablow 1974:I:2–I:3).

Some modern researchers have argued that by the end of the 1650s, the Illinois had moved permanently southward and westward into the interior, near to or perhaps west of the Mississippi River. Whether they had moved there en masse, or had actually been forced to resettle there and for what reasons is still not clear. According to some, these moves took place in response to escalating hostilities on the part of the Sioux and the Iroquois (Zitomersky 1994:77–79). Temple (1966:13), for example, speaks of Illinois migration across the Mississippi as a "retreat," the Illinois having been "forced to abandon their ancestral lands to the Iroquois beaver hunters." Charles Balesi (1992:22) argues that by this time the population of the Illinois had declined and they lacked "confidence," their influence in the region having declined significantly. He claims that they "migrated annually" to points west of the Mississippi to avoid complete devastation by the Iroquois.

Counter to the above argument, Jablow (1974:I:8–I:16) contends that the Illinois were indeed influential and populous during this time. He acknowledges that while they could have spent some time west of the Mississippi, he sees no evidence from contemporary accounts that the Iroquoian threat to the Illinois was actually imminent or that the Illinois should have been so frightened by them that they abandoned entirely their homelands east of the Mississippi and retreated across the river. Additional counterstatements are made by Callender (1978:678; see also Thwaites 1896–1901:59:161), Mary Elizabeth Good (1972:2), and Zitomersky (1994:79). However, while it is clear that although some of the Illinois are found back on the western shores of Lake Michigan trading and visiting in the late 1660s, after 1670 at least one major Illinois settlement (the Iliniwek Village, discussed below) was still located west of the Mississippi.

Documentary sources compiled by contemporary European observers of the Illinois provide reliable evidence that by 1673, various subgroups of Illinois occupied lands that stretched from above what is now Starved Rock, Illinois, to the Black River in Arkansas. A 1673–1674 map attributed to Jesuit missionary Father Jacques Marquette (Figure 5.2) illustrating his historic trip down the Mississippi River with French agent Louis Jolliet in spring of 1673 indicates the approximate geographical locations of Peoria, Moingwena, and Michigamea settlements along major western tributaries of the Mississippi, along with Kaskaskia and Maroa territories situated well into the Illinois Country to the east (Thwaites 1902:plate opposite p. 212). Over the course of their 1673 voyage, Jolliet and Marquette visited the Peoria and Kaskaskia settlements personally, reporting that the Peoria village cluster contained over 300 cabins, while the Kaskaskia Village on the Illinois River (now the Zimmerman site, La Salle County, Illinois) to north and east contained 74 (Thwaites 1896–1901:59:123). It would not be until the late 1670s that several of the Illinois subgroups coalesced at Kaskaskia, bringing the number of cabins there to 351 by 1677 (M. Brown 1975:1).

THE ILLINOIS AND THE ARRIVAL OF THE EUROPEANS

French penetration into the Western Great Lakes began in the second quarter of the seventeenth century with the arrival of Samuel de Champlain's agent, explorer Jean Nicolet near what is now Green Bay, Wisconsin, in 1634.[4] Their primary interests in native peoples of the region were in converting them to Christianity, promoting previously untapped sources of commerce, and in creating new military alliances. Pressures of Dutch and British trading and military competition below the St. Lawrence were undoubtedly spurring incursions into the midcontinent (Neill 1884:165, 172).

The French were also eager to investigate and exploit the natural resources purported to be in the region. Important among these were the "red copper" deposits (Kellogg 1968:345–349). Jacques Cartier had known of them since 1535 (Kellogg 1968:346). Jesuit Father Francesco Bressani (Thwaites 1896–1901:39:243) had written in 1653:

> There is a Copper ore, which is very pure, and which has no need of passing through the fire. . . . We have seen it in the hands of the Barbarians, but no one has visited the place.

During the 1660s, the French became actively interested (Moussette 2001: 323). Jesuit missionary Father Claude Allouez would report extensively on the significant copper deposits and mines of Lake Superior and its islands and of

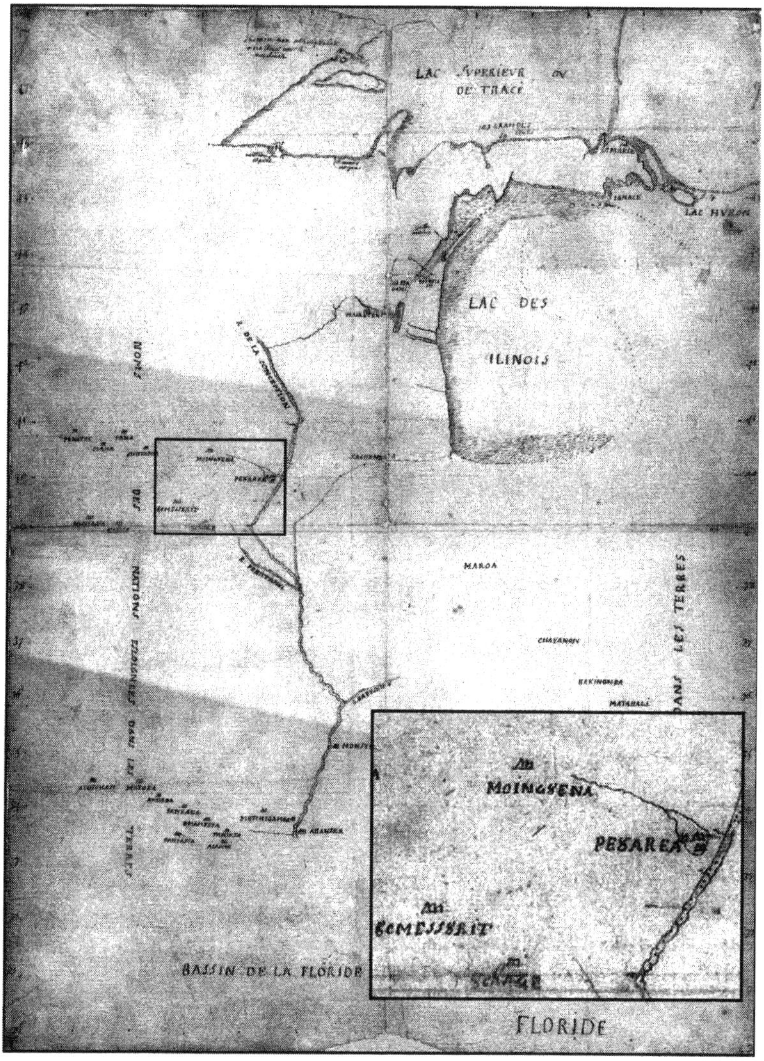

Figure 5.2 Marquette's 1673–1674 map of the Mississippi Valley. Inset features the three villages of "Peouarea." Archives de la Compagnie de Jésus, Saint-Jérôme, Québec.

the Jesuits' intention to investigate them (Thwaites 1896–1901:54:153–165). In 1667, French intendant Jean Talon sent an expedition there in search of them (Kellogg 1968:348–9; Thwaites 1896–1901:50:324). According to Helen Tanner (1987:36), the search for copper mines was the stated reason for annexing interior North America to New France in 1671.

Subsequently, and especially following the Huron diaspora of 1649, French explorers, traders, and missionaries ventured with increasing frequency from Québec and Montreal through the Straits of Mackinac into the lands west of Lake Michigan and south of Lake Superior. While home to several indigenous Algonquian and Siouan groups, including the Winnebago, Potawatomi, Menomini, and Chippewa, these lands were filling rapidly with remnant Algonquian and Iroquoian peoples who had recently fled from their previous homelands to the east in fear of the advancing Iroquois. Some, including the Huron, Petun, and Neutral, had already been vanquished by them (Mason 1986:381; Stone and Chaput 1978:602; White 1991:1, 11, 14). The region became a refugee center (White 1991:11, 14). Many displaced bands settled in contiguous villages or came together in large multiethnic village units wherein traditional social and political interactions were modified to accommodate these new settlement conditions (Hickerson 1971:176; White 1991:14–24).

The ever-expanding presence of Europeans in the Western Great Lakes and the growing trade in furs had undoubtedly brought native peoples of the interior to the Sault Ste. Marie (at the juncture of Lakes Huron and Superior) to trade from the 1640s. The Huron had become the primary "middlemen" (see endnotes 2 and 7, this chapter) in trade into the region; as a result, some Western Great Lakes peoples, including the Illinois, certainly gained knowledge of trade goods during this early period (Blair 1996:322; Mason 1997:306; Stone and Chaput 1978:602). While native/native exchange probably brought foreign goods into their lands, independent traders called *coureurs de bois* also circulated trade goods to native peoples long before officially sanctioned French contingents arrived (Brose 1971:56).

However, it was during the 1660s that the wave of European westward expansion intensified and "fur trade" activities began in earnest in the Upper Great Lakes (Kehoe 1981:240; Mason 1997; Stone and Chaput 1978:602). Once French victories over the Mohawk removed Iroquois middlemen from blocking access to the western fur trade and the government guaranteed prices for beaver pelts, the door opened for establishment of missions, forts, and trading posts (Emerson and Mansberger 1991:149–150; Stone and Chaput 1978:602). In 1665, the French established the trading post and mission of St. Esprit on the southern shore of Lake Superior at what is now Chequamegon Bay, Wisconsin (Tanner 1987:37). According to Jesuit missionary Father Claude Dablon, the Illinois were the first to visit St. Esprit (Thwaites 1896–1901:55:207). In the *Relation* of 1669–1671, Father Dablon reports that the Illinois "repair to this place from time to time in great numbers, *as Merchants,* to carry away hatchets and kettles, guns, and other articles that they need"

(Thwaites 1896–1901:54:167, emphasis mine). From his post at the same mission, Father Jacques Marquette later notes that the Illinois were warriors and slave traders, trading with the Ottawa for "Muskets, Powder, Kettles, Hatchets, and Knives" (Thwaites 1896–1901:54:191).

It is clear that by 1670, the Illinois are interacting actively with the French and with native purveyors in what appears to have been an enterprising effort to gain and keep access to trade goods. Indeed, they seem to have been interested in making alliances with the French and themselves becoming intermediaries in the trade to the interior. Once another French mission and trading establishment, St. Francis Xavier, was founded at Green Bay, Wisconsin, in 1669, some Illinois are found settled however permanently near that mission as well (Tanner 1987:37; Thwaites 1896–1901:55:13–15).

Foreign entanglements grew even deeper for the Illinois as trading parties received religious instruction from the Jesuits during their sojourns at St. Esprit. It is there in 1666 that Father Allouez recounts meeting and proselytizing 80 Illinois who had come to the post to trade (Thwaites 1896–1901: 51:49). In the *Relation* of 1669–1671, Father Dablon reports that the Illinois were anxious for further evangelization, and that they had repeatedly expressed their desire to be ministered to more regularly in their own country. Indeed, in the same relation, Father Marquette, also writing from St. Esprit, calls the Illinois "lost sheep" who "cry so loudly that one hastens to rescue them from the jaws of the Wolf." To that end, he discussed plans to accompany the Illinois to their own country the following year to establish a mission among them (Thwaites 1896–1901:54:187–191).

THE FRENCH IN THE ILLINOIS COUNTRY—THE MARQUETTE AND JOLLIET EXPEDITION OF 1673

As French economic and missionization efforts expanded in the Western Great Lakes, so did their desire to know more about the resources and the native peoples of the interior. In 1673, Canadian intendant Jean Talon chose fur trader Louis Jolliet to lead an exploratory mission to find and explore the Mississippi River. Jolliet's mission was to find new natural resources to exploit. He was to be accompanied by Father Jacques Marquette, who would be scouting a location for the proposed mission to the Illinois. Jolliet's account of the expedition was lost in a canoe accident in 1674; however, Marquette's journal[5] still survives (Thwaites 1896–1901:59:113–137). It relates their encounters with at least two, possibly three different Illinois subgroups at four different locales along the route. A map also attributed to Marquette shows their locations (Figure 5.2).

Marquette's extended narrative recounting the explorers' visit to one large settlement, called "Peouarea" (or "Pe8area")[6] by the Illinois marks the first detailed ethnographic account of the Illinois. It describes their settlement as consisting of three adjacent, nearby villages located a short distance up a major western tributary of the Mississippi River (Figure 5.2, inset). One village was on the floodplain and two more stretched along the top of a hill upriver from the first. The journal recounts their welcome into one of these villages by individuals who identified themselves as Illinois and the several ceremonies that were held over the course of their visit.

According to Marquette, the Illinois of Peouarea recognized the visitors as "frenchmen," and Marquette as a "black gown." In turn, Marquette identified them as allies, as they were dressed in cloth (Thwaites 1896–1901:59:15–117). They also possessed guns, which Marquette claimed they used to scare, rather than to kill, their enemies. He further adds that the Illinois had obtained them through purchase from "our savage allies who Trade with our french" (Thwaites 1896–1901:59:127). As was native custom, Marquette presented them with gifts as important elements of ritualized exchange (White 1991:36).

The narrative also yields important clues as to one strategic means through which the Illinois sought to secure a foothold in the "fur" trade. During their visit, the Illinois "captain" presented their foreign visitors with a young slave as a valued gift. While it is unclear whether the slave was Illinois, the narrative describes how the process worked. According to Marquette (Thwaites 1896–1901:59:127), the Illinois,

> make themselves dreaded by the Distant tribes to the south and west, whither they go to procure Slaves; these they barter, selling them at a high price to other Nations, in exchange for other Wares. Those very distant Savages against whom they war have no Knowledge of Europeans; neither do they know anything of iron, or of Copper, and they have only stone Knives.

It is not at all surprising to learn that the Illinois traded slaves for European trade goods. One of the reasons for trading slaves instead of (or in addition to) pelts may have been that beaver were less abundant below the Wisconsin River (Jablow 1974:I:10; White 1991:143; Wiegers 1985:75–77). In the Mississippi Valley, captive/slave taking practices predate European contact, and both contemporary and modern sources attest to its widespread occurrence in the early contact period for a variety of purposes (Hauser 1973; Pease and Werner 1934:381–382; Wiegers 1985:75–77). These reasons notwithstanding, the Ottawa[7] appear to have been willing trading partners in this enterprise.

Armed with a calumet of peace given to them by the Illinois, Marquette and Jolliet left the "Peouarea" village. The small entourage proceeded southward, passing only long enough to present gifts at a Michigamea village (Thwaites 1896–1901:59:151). They continued until they reached the mouth of the Arkansas River. Realizing then that the Mississippi River did actually flow into the Gulf of Mexico, they decided to turn back in fear of attacks by the Spanish. They returned to St. Francis Xavier mission via the Illinois River, which took them past the Kaskaskia village (containing 74 cabins), near Utica and Starved Rock, Illinois (Thwaites 1896–1901:59:161). On the return trip, Marquette also mentioned (Thwaites 1896–1901:59:163) that "we passed through the Ilinois of Peouarea, and during three days I preached the faith in all their Cabins." This is problematic, however, because as Marquette described their return route, it would not have taken them past the Peouarea village he had previously described. One scholar postulates that the encounter may have actually taken place at a smaller settlement of Peoria at Lake Peoria, near modern Peoria, Illinois (Franke 1995:11).

Jolliet and Marquette's historic voyage of 1673 was a watershed event in the history of the North American interior. It virtually opened the Upper Mississippi Valley and the Illinois Country to direct contact by the French descending from the north. The next decade would be a period of tumultuous activity and movement for the Illinois. This was largely due to intensifying native/native political relations compounded by the French trying to strengthen native alliances and establish their own foothold in native territories.[8] In 1675, Father Marquette returned to the Illinois Country and founded the Immaculate Conception mission at the Kaskaskia Village (Donnelly 1968:251). He died soon thereafter, and was replaced by Father Allouez in 1677. Soon after 1673, the Kaskaskia Village grew substantially and housed eight groups of Illinois rather than only the Kaskaskia. Allouez reported that in 1677, the village contained 351 cabins as opposed to 74 in 1673. According to Temple, this increase likely came as a result of renewed threats from the Iroquois to restart their war with the Illinois, whom the Iroquois had long regarded as a stubborn yet powerful enemy and as a serious threat to their fur trade hegemony. The Kaskaskia called other Illinois to come together at Starved

Rock to more effectively repel Iroquois raiders (Temple 1966:20). By 1679, the village had increased to between 400–500 cabins (Temple 1966:21). Meanwhile, the Illinois had also renewed hostilities with the Fox who lived near Green Bay (Temple 1966:20).

In 1678, the governor general of New France commissioned René-Robert Cavelier, Sieur de La Salle to undertake an exploratory expedition and to establish trading posts in the Illinois Country (Hall 1991:14). In 1680, descending the Illinois River, he encountered a large Illinois village at Lake Peoria, or Pimitoui. Fort Crèvecoeur was constructed nearby. Later that year he left Crèvecoeur, which the remainder of his party deserted shortly thereafter. However, one of his assistants, Henri de Tonti had left beforehand to go upriver and explore Starved Rock. In September 1680, the Iroquois attacked the Illinois and Tonti at the Kaskaskia Village. Both were ill equipped to fend off the attack or to broker peace and fled the location.[9] The Iroquois destroyed the village.

Thereafter, Illinois hostilities with the Iroquois intensified, only to be complicated by ever-present, now escalating strife with the Sioux. Intra- and intergroup alliances, largely based on gaining protection from enemies and keeping access to the trade, were made and broken (White 1991:29). Retreats and outmigrations to remote villages, subtribal mergings, further depopulation, surrender and taking of captives, even murder of independent Canadian traders appear to be among the consequences of these activities (Hauser 1973:22–23; White 1991: 29, 31). At the same time, the interests and motivations of the French to keep the Illinois allied to them and to prevent peace with the Iroquois (who were allied with the English) cannot be ignored (Beaulieu and Viau 2001:14–15; Temple 1966:21–22; White 1991:30).

In 1683, threats of new attacks by the Iroquois and a wish "to establish an empire in the Illinois Country for fur trade" led La Salle to act aggressively in the interests of New France (Temple 1966:26–27). He attempted to ally many of the groups of the region, promising to build a strong native/French alliance and to provide protection from the Iroquois. A substantial number of groups, including several subtribes of Illinois and Miami, plus Mascouten, Shawnee, and Wea, coalesced around a new fort he built at Starved Rock called Fort St. Louis. The alliance exemplified the way in which the French dealt with native peoples of the Western Great Lakes during this period. Achieved out of native fear, intertribal conflict, and a desire to intensify trade, the strategy required diplomacy, mediation, and reconciliation without coercion (White 1991:24–30). Despite new raiding and a promise not to leave any Illinois alive, the Iroquois were not successful in immediately destroying the alliance. It did loosen by 1689 as groups moved away. The fort was abandoned by 1692 (Hall 1991:14; Temple 1966:29). Peace with the Iroquois would not be achieved until 1701 (White 1991:49).

LINKING THE ETHNOHISTORIC RECORD TO THE ARCHAEOLOGICAL RECORD— THE HAAS/HAGERMAN SITE

Searching for the Late Protohistoric/Early Historic Illinois

The archaeological record has significant potential to shed light on these events and conjunctures among the French and the various native entities present in the region. However, as mentioned in the introduction to this work and to this chapter, in Western Great Lakes protohistory, attempts to identify individual protohistoric ethnic groups and to link them confidently to particular archaeological manifestations has been an enduring problem for archaeologists (Brose 1971; J. Brown 1990a; Emerson and Brown 1992; C. Mason 1976, 1997; R. Mason 1976). As archaeologists well recognize, until reliable and sustainable linkages are made, many questions concerning native/native and native/European sociopolitical and economic interactions cannot be addressed at a culture-specific level. This section outlines how the Haas/Hagerman site, in Clark County, northeastern Missouri, was linked securely to the "Peouarea" village of the Marquette and Jolliet expedition, hence to the protohistoric/early historic Illinois.

Historically, attempts to associate a distinctive late prehistoric material culture assemblage with the Illinois had begun with Kenneth Orr's 1947 work at the Zimmerman site (the site of the Grand Village of the Kaskaskia) across from Starved Rock, Illinois (J. Brown 1961, 1990a:155). The Zimmerman site is known to have been the major summer village of the Kaskaskia in the early 1670s and the location where several other Illinois and native groups, including the Peoria, ultimately came together between 1675 and 1680 and again in 1683. Inquiry there has focused on sorting out the numerous complexes excavated at the Zimmerman site and associating particular ceramic styles found within them with known historic groups. These styles would serve as ethnic markers that could then be used (ideally) to trace these ethnic groups back through protohistory and perhaps even into prehistory (Emerson and Brown 1992; Walthall 1992:155–157). Among other complexes identified at Zimmerman, a Danner component in association with Danner series ceramics was defined in 1961 (J. Brown 1961:21, 74–76) and was elaborated upon by Margaret K. Brown (1975) after her own work there. After years of deliberation on the problem, the Danner complex emerged as the most likely candidate for association with the historic Illinois (J. Brown 1990a). However, connections remained inconclusive primarily due to the complex historical and archaeological conditions at the Zimmerman site that precluded making unequivocal ethnic assignments.

Archaeologists have long been aware of the difficulties in linking material culture assemblages with particular ethnic groups in order to trace them through time and space (Maceachern 1998:111). Some have questioned the legitimacy and/or utility of such a pursuit in the Western Great Lakes considering the tumultuous conditions prevailing there throughout the early contact period (Brose 1971; J. Brown 1990a:158; Emerson and Brown 1992:104–105; R. Mason 1976). Additionally, archaeologists also have become increasingly wary of using particular elements of material culture assemblages, namely ceramic styles, as reliable indicators of ethnicity. They have been particularly critical of the practice when assignments are based solely on the appearance and distribution of particular decorative styles (Cordell and Yannie 1991; Gibbon 1995; Stark 1998).

Isolating the Protohistoric Illinois

The Haas/Hagerman site, Clark County, Missouri (23CK116), was identified as a large historic contact period village during a 1984 water line survey when archaeologists (Walters and Boyd 1986:35, 49–50) discovered material evidence of European contact such as glass beads, copper sheet, and iron knife fragments among otherwise late protohistoric aboriginal cultural remains. Located in the lower Des Moines River valley approximately 8.5 miles upstream from its confluence with the Mississippi, the site is situated on a high glacial outwash sand terrace about 8 meters above, and 2.3 kilometers west of the river. It stretches in linear fashion along the leading edge of the terrace for approximately 2,100 meters north-south and about 200 meters east-west (Grantham 1993). A secondary tributary stream valley transecting the site serves to divide it into two sections—the area north of the stream is called the Haas (formerly Haas-Eckles) site and the portion south of the stream is referred to as the Hagerman (formerly Hagerman-Parsons) site.

From the time of its discovery, archaeologists recognized the Haas/Hagerman site's potential links to descriptions of the "Peouarea" village location where

Father Jacques Marquette and explorer Louis Jolliet encountered the Illinois in 1673 (Walters and Boyd 1986:35). Despite the fact that Marquette's description and also his map of the Mississippi River on which he placed "Peouarea" still survive, the actual village site had never been located on the ground.[10] Bringing historical, cartographic, and archaeological data to bear on the question, Larry Grantham (1993) has argued convincingly that the site was occupied during the mid-seventeenth century by groups of Algonquians who clearly had access to trade goods. Using the presence of Danner ceramics as a major line of evidence, Grantham further argued that these Algonquians likely represented one major manifestation of a western migration of Illinois across the Mississippi River that is known to have taken place during the mid-seventeenth century. He (1993:14, 18) concluded that the Haas/Hagerman site was the meeting place of Jolliet, Marquette, and the Illinois.

Fortunately, connecting the Illinois with the Haas/Hagerman site did not have to rest on ceramic style alone. Geographical, historical, and archaeological evidence supports the site's association with the Illinois. The Haas/Hagerman site is a clear (and extremely rare) example of what Ronald Mason (1976:351) has termed *site unit ethnicity*—that is, "a (hopefully) single-component site which has been located rigorously in time and space by unambiguous historical sources and which yields appropriate native and/or European artifacts."

The Haas/Hagerman site presents a situation in which historical sources approximate the location, size, and configuration of the site, identify its occupants as Illinois, and provide a date in which the site was known to have been occupied. Archaeological evidence has continued to corroborate the documentary record, site data revealing the "appropriate" or "likely" suite of early trade goods and intact native industries. To date, two components, a diffuse, poorly defined Woodland component, and at least one late protohistoric/early historic component, are known at the site (see Ehrhardt 2004; Grantham 1993). Calibrated radiocarbon dates from the latter occupation fall within a temporal range from about 1640 to the early 1680s, straddling the threshold of the historic period (see Grantham 1994). Danner ceramics clearly dominate the component's ceramics industry, and trade goods of types consistent with the time have been recovered (Grantham 1994; Ehrhardt 2004; Ehrhardt and Conrad 1994). Thus, while the multiethnic, fluid nature of native villages is well known for the Protohistoric period in the Western Great Lakes, the early contact period component at Haas/Hagerman represents at the very least the ethnically undifferentiated Illinois.

HISTORY AND RESULTS OF ARCHAEOLOGICAL INVESTIGATIONS AT THE HAAS/HAGERMAN SITE

The Haas and Hagerman sites have both been investigated archaeologically since 1993 by the Missouri Department of Natural Resources, the Upper Mississippi Valley Archaeological Foundation/Western Illinois University, and the University of Illinois. Research has taken place in areas A1, B1, C, D, and F1 and F2 (see Figure 3.1). Multiple strategies, including controlled surface collections, geophysical techniques, and intensive hand excavation, have been utilized in different areas of the site depending on individual project research designs.

Results of these investigations have yielded important information about mid-seventeenth-century Illinois community layout, domestic life, subsistence economy, technology, and early European material influence. The village is laid out in linear fashion on a north-south axis along the leading edge of the terrace. It extends landward only to a depth of about 80–100 meters west of its eastern edge. Such a linear settlement plan is consistent with that found at the roughly contemporaneous Zimmerman site (J. Brown 1961:14). In addition, Marquette's (Thwaites 1896–1901:59:123) description of the hilltop village(s) as being made up of rows of longhouses (cabins) separated by roads is confirmed by resistivity. These results indicate that in the area surveyed, the village is three rows of structures deep (Grantham 1996). How uniform this pattern is over the entire north-south extent of the site is unknown, however, it is clear from surface survey that it is likely to extend through areas C, D, and F1 and F2. House spacing is not as regular in Area B1 nor is it as uniform as in Area C. Village depth quickly diminishes in Area A1.

At least parts of the village were also found to have been fortified on the landward side. A stockade was identified in Area B1 through resistivity and was verified by hand excavation. The revealed section appears to have been made up of vertical logs. It had one bastion (tower) and one opening, and was rebuilt at least once. Despite the site's relatively remote location, the fact that it was fortified is not altogether surprising since the Illinois were in almost constant threat from the Sioux and the Iroquois from the 1650s. Their nearest Siouan neighbors known archaeologically may well have been the Missouri, who inhabited the Utz site, located on the Missouri River in Saline County, Missouri, 185 miles from Iliniwek (Henning 1970; Wedel 1959:39). Other Chiwere Siouan groups occupied the McKinney site, only 61 miles away. This site is thought to have been abandoned by 1645, however (Tiffany 1988). What the Illinois moving into the area may have had to do with McKinney's abandonment is unknown.

A variety of structure types has also been defined at the Haas/Hagerman site. One complete longhouse with five large central posts and an exit pathway, as well as portions of two additional dwellings of this type have been excavated in Area C. They are of unequal size. In addition, a small oval structure and a fairly large round structure, both of unknown function, have also been explored completely. The oval feature shows evidence of repeated rebuilding.

Storage/refuse pit features are plentiful at the site. They are distributed both inside and outside of houses. Often overlapping, pits vary in size, shape, depth, and density of cultural remains. Only one appears to have been lined with matting. Their contents most often consist of village or house debris. Pit fill is typically homogeneous, although a small proportion was stratified. Most contain some sort of trade item, indicating that they were filled at a time when the Illinois had access to trade goods.

Importantly for any evaluation of early contexts of material culture continuity and change during this period, the artifactual assemblage from the site reveals intact native industries in stone, ceramics, and osseous materials. These artifacts were recovered from both pit feature fill and the excavation levels above them. Chipped and ground stone artifacts and wastage comprise by far the single most plentiful raw material category. Chipped stone artifacts include small, Madison type triangular projectile points, unifacially worked end and side scrapers (some in combination), and various types of bifaces. Small flake graver/piercers or "micropiercers" are also noted in the industry. Abrading stones, hammer stones, nutting stones, anvils, manos, and metates comprise the ground stone industry.

Several pieces of worked catlinite are featured in this industry as well. Two triangular sawed, snapped, and ground catlinite blanks, perhaps meant to be used as perforated janglers, were recovered as were two pipe (calumet?) fragments.

The ceramics from the Haas/Hagerman site are dominated by Danner ware (Figure 5.3). This ware (discussed above) has now been associated firmly (but potentially not exclusively) with the protohistoric/early historic Illinois due to its presence in such significant proportion at this site. Grantham has further identified a new type, Haas, within the ware category (Figure 5.3 c)

Figure 5.3 Danner ware ceramics from the Iliniwek Village: (a) rim sherds with pinched out, and punctated (left) or incised (center and right) rim strips; (b) rim sherds with lug (left) or strap (right) handles. Sherd on the left has a tab on either side of the lug; sherd on right has a single tab over the strap; (c) Haas rim sherd with finger impressions at the shoulder.

(Grantham 1996; Grantham and Ehrhardt 2000). Of the types identified, Danner Cordmarked dominates. Keating Cordmarked is represented by only two rim sherds. The industry also contains about 1–2 percent Allamakee Trailed ceramics. Allamakee Trailed is the single type of Oneota tradition ceramic that has been clearly and unambiguously linked with several closely related protohistoric/early historic Chiwere-speaking Siouan peoples (Hall 1993:26; Wedel 1959:39). A very small number of unidentified, presumably exotic types have also been recovered.

Osseous implements, ornaments, and worked raw materials are also present. Forms include bison scapula hoes, an awl, a long mat-weaving needle, antler handles, and several fragments of antler flakers. Other bone objects, including a few (less than six) bone beads and a single bone tubular counter occur in the industry. A cache of curated, worked antler has also been recovered, as has a single piece of incised conch.

Pit matrices also contain archaeobotanical and faunal evidence essential to understanding the potentially changing roles of plant and animal resources in the diet and subsistence economies of peoples of the period (Walz 1996). Tropical cultigens consist principally of corn with lesser amounts of beans and squash. Native cultigens are limited; they are represented by little barley, chenopodium, tobacco, sunflower, and knotweed. Watermelon is the only confirmed Old World domesticate recovered. Wild plant seeds include but are not limited to wild rice and various types of berries, grasses, and other fruits. Hickory, black walnut, and acorns comprise the major types of nut remains represented (Hollinger 2001).

Faunal remains are abundant and reflect a wide variety of exploited terrestrial and aquatic species. Patterns at Haas/Hagerman vary from those at Zimmerman based on what are interpreted to be differences in local exploitation practices (Martin, Richmond, and Brand 2003). Analyses of more than 83,000 animal remains from several houses at the Haas/Hagerman site demonstrate that mammal bones predominate (68 percent of the specimens), with white-tailed deer, dog (canid), beaver, raccoon, bison, tree squirrel, black bear, elk (or wapiti), muskrat, cottontail, bobcat, and mountain lion appearing in the assemblage (in decreasing order of specimen frequency). Fish specimens account for 17 percent of the assemblage. Remains of large fish such as gar and freshwater drum are most frequent. Other fish include channel catfish, flathead catfish, bullhead, shovelnose sturgeon, and various suckers. Freshwater mussels also occur. Birds, however, are underrepresented. These include wild turkey, ducks, pied-billed grebe, bald eagle, American crow, passenger pigeon, mourning dove, as well as small perching birds.

These findings broadly support contemporary ethnohistoric documentation that characterizes late-seventeenth-century native life in the Illinois Country as semisedentary with bison hunting expeditions typical of the summer and early winter (Emerson and Brown 1992). The presence of corn, beans, and squash argues for a seasonal horticultural economy, while bone implements and remains attest to bison utilization. Heavy local exploitation of deer and fish resources undoubtedly provided for a diversified resource base. However, deer also provided skins that may have served the Illinois (as they served others farther to the south) as commodities in the trade (Braund 1993:61; White 1983:34–35). The presence of beaver in the faunal assemblage is also provocative. Whether its occurrence, in even moderate amounts, reflects shifts in hunting patterns to accommodate the fur trade rather than subsistence needs is unknown.

The presence of dog remains in the Haas/Hagerman faunal assemblage deserves further comment. Numerous contemporary observers chronicle aboriginal use of dogs in supplication rituals, but principally in feasting (Blair 1996:53, 60). Marquette makes specific reference to his participation in a feast at Peouarea in which the Illinois offered their guests cooked dog meat as a sign of the high esteem in which the Illinois held them (Thwaites 1896–1901:59:123). Dogs were also a particularly integral part of Illinois war feasts (Pease and Werner 1934:377).

The significance of trade goods appearing at Haas/Hagerman in the form of glass beads and copper-base metals, lead, and iron in notable quantities (for the time) cannot be overemphasized. While it is not the earliest site in the Illinois Country to yield trade goods (J. Brown 1990b:236; Lurie 1998),[11] the presence of these materials so deep in the interior by about 1640 or even earlier clearly signals early interaction either with Europeans or with other native peoples who had access to trade goods and moved them through native networks. Penelope Drooker (1996a:145) maintains that many of the glass and metal items she reports on from Late Fort Ancient contexts at Madisonville in the central Ohio River Valley in Ohio may have been available ca. 1600, and that as early as the *late sixteenth century* "at least three major interaction networks intersected at Madisonville." Her findings certainly attest to the speed at which trade goods potentially moved through native hands.

Certainly, trade goods had been available to native peoples of the Southeast as early as 1521, and became more prevalent after the de Soto *entrada* of 1539–43. However, as Marvin T. Smith (1987:25) says, "until the founding of Charles Towne in 1670, there was no regular trade in furs or deerskins with the interior of the Southeast, and European goods must have been fairly rare." Marquette relates that trade goods (of undisclosed type) had reached unnamed groups he encountered in the lower Mississippi Valley in 1673. In his words, (Thwaites 1896–1901:59:149) these peoples had,

> guns, hatchets, hoes, Knives, beads, and flasks of double glass, in which they put Their powder. They wear Their hair long, and tattoo their bodies after the hiroquois fashion. . . . They assured us . . . that they bought

cloth and all other goods from the Europeans who lived to The east, that these Europeans had rosaries and pictures; that they played upon Instruments; that some of them looked Like me.

The source or sources for trade goods of which Marquette speaks is unknown. Some peoples of the lower Mississippi Valley had access to guns and trade goods via the English, who were operating out of Carolina (Sabo 2000:187; see also Gallay 2002 for English/native trade activities in the coastal Southeast prior to 1673). By the end of the seventeenth century, the English and their native allies penetrated the southeastern interior, raiding for native slaves (Gallay 2002; White 1983:35). Marquette's statement would even less likely refer to the Spanish who established early missions on the southeastern coast (McEwan 2000). However, James Brown and Robert Sasso (2001:220) have recently pointed out potential influences from the Spanish in the Southwest. Lastly, the French did not officially arrive on the Gulf Coast until 1699 (White 1983:35). They had tried unsuccessfully to establish a colony in La Florida between 1562 and 1565, but failed several times. With the Spanish eager to rout them, they finally abandoned their efforts (Milanich 1999:78–81).

The range of trade goods at the Haas/Hagerman site is quite narrow. The inventory consists of metal artifacts and scrap, glass beads, and a Jesuit ring. Metal objects, stock, and scrap are made of copper-base metal, iron, formed and molten lead, and silver. Except for a twisted brass butt plate, a ferrule, and several lead musket balls, few gun parts or gun paraphernalia were recovered. Iron artifacts include axes and axe fragments, clasp, butcher, and utility knife blades, butcher and case knife blade fragments, knife handle rivets, kettle bail fragments, an iron coil, one piece of bar stock, and pieces of unidentified objects. Many artifacts were oxidized beyond recognition.

Copper-base trade metal occurs in reworked form and as wastage or "scrap." Finished and unfinished artifacts were recovered. These articles were primarily ornamental or personal adornment items including beads, coils, bracelets, rings, pendants, tinkling cones, and clips. Unfinished cut blanks, tubing, wire, and scrap metal fragments are also found. Detailed descriptions of these artifacts will be presented in chapter 6. Linda Longoria (1998:136) reports four copper-base metal projectile points from private collections. None was excavated under controlled conditions.

Glass beads number about 3,000. They range in size and type. While monochrome blue and black beads were most common (comprising about

85 percent of the total beads recovered), white-cored, brick red specimens as well as white, brick red, red, and colorless beads are also noted. Black and navy blue beads with white stripes also occur. One white-striped navy blue specimen is drilled into laterally. Surprisingly, four seven-color star chevron bead fragments were recovered; some of the broken fragments were ground smooth and appear to have been curated.

IDENTIFYING CONTEXTS OF CHANGE

Taken together, the documentary, ethnohistoric, and archaeological records paint a complex and multilayered picture of the mid-seventeenth-century Illinois during the earliest years of their contacts with Europeans, their ideologies, and their merchandise. Patterns of escalating intrusion and eventual colonization, locational and settlement upheaval, depopulation, resource exploitation, political manipulation, and material, social, and ideological transformation were only beginning to take shape as the Illinois and Europeans met during this period in what Richard White (1991) has called "the middle ground" of interethnic relations. These activities took place on a historical stage of "conjuncture" in which societal structures are potentially altered and renegotiated to meet new conditions according to the values, ideas, motivations, and intentions that each player brought to the event or unfolding process (after Sahlins 1981:35–37).

The timing and tempos of these points of change were not synchronous nor did they have equally devastating effects. Processes are clearly in motion, however, and the Illinois are shown to be active agents driving them. For the Illinois, like many other native groups, the material culture realm is one of the earliest contexts in which changes are clearly documented. Desire for trade goods and an interest in expanding the frontier trade economy into the south and west at the expense of their neighbors were both strong factors that undoubtedly pushed the Illinois toward ever-increasing interaction with Europeans. Other clearly documented domains of response are remote village location, village fortification, and an apparent interest in increased involvement with Europeans at least through evangelism and exchange of goods. One might also hypothesize that an increase in raiding (for slaves) heightened hostilities with remote neighbors even as relations with the Sioux and Iroquois continued to deteriorate. Shifts in resource procurement to meet the economic demands of foreign exchange may also have been an increasingly influential variable that came into play at this time.[12]

6 / From Kettle Sheet to Ornament

Artifact Forms, Production, and Use

THE COPPER-BASE METAL INDUSTRY

This phase of the materials analysis identifies, classifies, and describes the range of artifacts in the Iliniwek Village copper-base[1] metalworking industry and places these objects within known and potential contexts of use during the late protohistoric/early historic periods. Consisting of a rather narrow array of decorative, ornamental objects rather than utilitarian "tools," the 806-artifact sample breaks down into four major artifact categories: 1) finished artifacts (complete and fragmentary); 2) unfinished/partially processed artifacts (complete and fragmentary); 3) often heavily manipulated, but apparently rejected "wastage" commonly called "scrap"; and 4) a small miscellaneous "other" category (Table 6.1). As Table 6.1 shows, finished artifacts outnumber unfinished ones. A significant portion of the industry (27 percent) is made up of material that has been partially reworked in preparation for making artifacts. Fragments of finished and unfinished artifacts and scrap, or material that may have been heavily manipulated but which was apparently rejected without being shaped into any recognizable form, each constitute equal portions (17 percent) of the assemblage.

Some general observations are noteworthy at this early point in the investigation. First, the industry contains finished, partially finished, and discarded materials. These artifacts are found in all excavated areas of the site, indicating that the Illinois were indeed working with copper-base materials at the village. If they had been receiving only finished artifacts made specifically for

Table 6.1 Iliniwek Village Copper-Base Metal Industry

Finished Artifacts			
Type	Complete	Fragments	Total
Beads			
(small rolled and tubular)	140	0*	140
Clips	128	24*	152
Tinkling Cones	21	1	22
Rings (hollow tubing)	5	0	5
Coils (hollow tubing)	4	0	4
Beads, spiral (strip)	3	0	3
Tubes, hollow	3	1	4
Awls	1	0	1
Pendants (triangular)	1	4	5
Bracelets (hollow tubing)	0	1	1
Bell, hawk	0	1	1
Total	**306**	**32**	**338**

Unfinished/Partially Processed Artifacts			
Type	Complete	Fragments	Total
Blanks	171	108	279
Tubing segments, B-shaped	34	0	34
Tubing segments, e-shaped	10	0	10
Tubing segments, o-shaped	1	0	1
Wire, solid	2	0	2
Total	**218**	**108**	**326**
Scrap	**139**	**n/a**	**139**
Other			
Kettle	0	1	1
Small, bipointed object	1	0	1
Nodule	1	0	1
Total	**2**	**1**	**3**
TOTAL	**665**	**141**	**806**

* since it was not possible to distinguish curved bead fragments from curved clip fragments, the "clip fragments" category is likely also to contain bead fragments

the Indian trade from European traders or from native middlemen, it is not likely that scrap and unfinished/partially processed materials would occur. This is not to say that the Illinois *never* received finished objects; it merely provides strong evidence that they were fabricating them locally in what appears to have been an ongoing process of artifact manufacture.

Second, *no* complete kettles and *very* few kettle parts other than sheet are found in this industry. Only one riveted lug, which had been cut away from the body of a kettle, was recovered. There are also iron bails (handles) and iron bail or rim fragments in the site assemblage. It is tempting to assume that the virtual absence of broken, or "used up" and discarded kettles and the small number of kettle parts recovered means that all copper-base metal parts of the kettle were reworked. However, it may also mean that the Illinois only rarely received complete kettles, obtaining kettle sheet in smaller pieces instead. Both Penelope Drooker (1996b:172) and William Fitzgerald (1990:545–546) draw attention to the "gift packs" of multiple pieces of cut copper-base sheet metal bound with fiber cord found at the Hood site (Neutral) as a means through which sheet material changed native hands from the early sixteenth century. It may also be that any whole kettles the Illinois received may have ended up in mortuary contexts or were deposited in areas of the site that have not yet been investigated.

Third, the reuse of the metal in this sample is exclusively for personal adornment items, not utilitarian (technomic) objects. This points to a non-need, non-subsistence based motivation for procurement and use of purportedly "higher quality" or "more efficient" copper-base metal material for weapons and implements by the Illinois (see Hamell 1983:18). Considering the kinds of ornamental items that have been found at other early contact sites in the region, a limited, but overall predictable, assemblage of personal adornment items is present at a relatively early time in the midcontinent.

Fourth, the metal sheet from which the artifacts were fashioned is generally flat, smooth, and of uniform thickness of between .2–.4 mm; its surface topography does not display the characteristic peening marks or spinning striations apparent on the much thicker kettle metal found on native sites elsewhere to the east. These characteristics render the metal, whether smelted copper or brass, relatively easy to shear, bend, and shape. Whether this means that kettles made for trade were being manufactured of thinner material during this period[2] or that the native source for them was different is not known.

Fifth, the condition of the metal varies. All artifacts are corroded to some degree. Generally, attack is moderate. Importantly, although individual artifacts appear quite heavily and/or differentially corroded, their original shapes are still preserved in the minerals of the corrosion product (after Smith 1981a: 78). Stress and/or corrosion cracking penetrates the bodies of some of the artifacts, especially in areas where deformation has occurred. Scrap and tubing are particularly affected by corrosion. This is likely due to the amount of manipulation to which these types were subjected before abandonment. The margins of especially thin specimens are particularly friable.

Lastly, the amount of material that is unfinished, partially processed, or thrown away is striking. It raises two important questions regarding the use value and symbolic meaning of these materials: 1) if foreign copper-base metal is so highly valued, why are artifacts discarded in refuse pits before they are finished, or abandoned as "wastage" in amounts more than adequate to manufacture personal adornment items, and 2) even if the Illinois were not using these materials themselves, would they not recognize their worth (no matter how fragmentary) as a commodity for trade into the interior, given that they were known to have been capturing slaves precisely for that purpose? These questions are treated in subsequent chapters.

Keeping these contexts of procurement, manufacture, and use in mind, results of the technometric and microscopic investigation of object types within the major finished, unfinished, and scrap categories follow. Findings from the documentary record and previous archaeological research concerning these types of materials, their distribution, and use among the Illinois and elsewhere are woven into the discussion (see Figure 6.1 for sites and locales mentioned in the text).

FINISHED ARTIFACTS

Small Rolled and Tubular Beads

According to S. Terry Childs (1994:235), rolled ornamental beads made of hammered sheet are the most prevalent native copper objects to appear in the northeastern United States and southern Canada from the Late Archaic through the Late Woodland period. Prehistorically, these copper items often occurred alongside formal and functional counterparts made of shell and bone. When European-introduced copper-base metal became available, native peoples of the protohistoric adopted it to produce these same forms known prehistorically. Cylindrical beads made of native copper as well as smelted copper and brass are found as personal adornment items on protohistoric and early contact period sites in the Northeast (Drooker 1996b:158; Fox

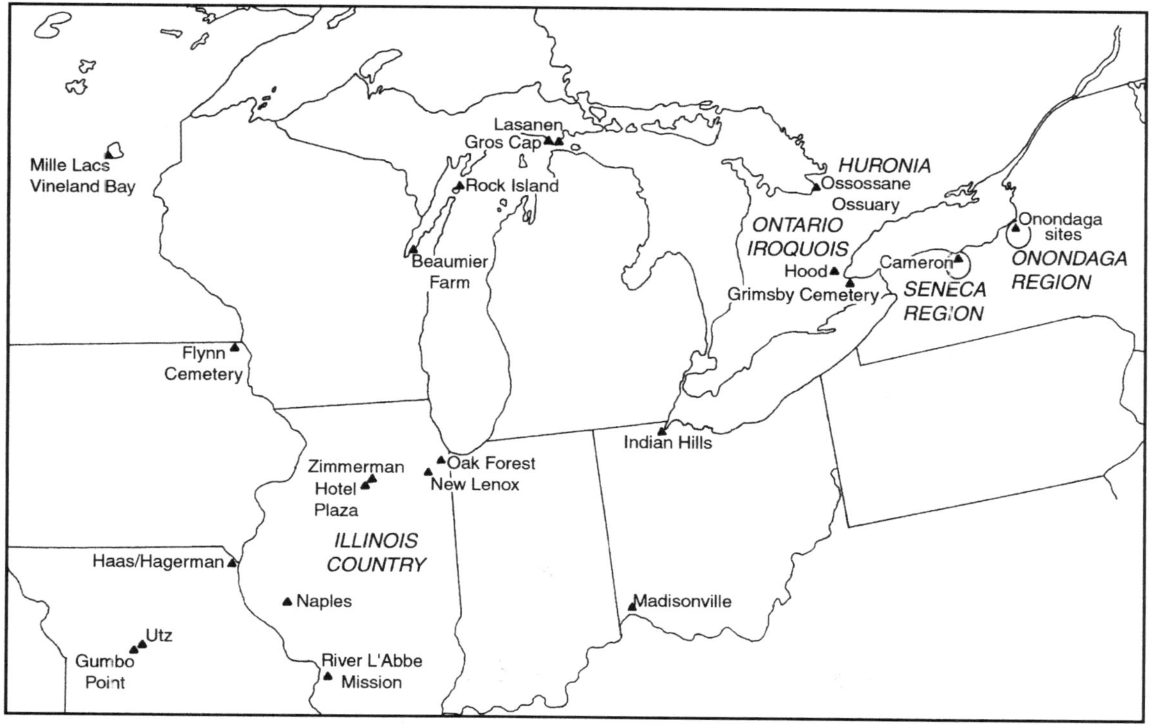

Figure 6.1 Early contact period Great Lakes and Mississippi Valley sites mentioned in chapter 6.

et al. 1995:288). However, as mentioned in chapter 4, native-made beads of imported, smelted copper-base metals gradually replaced native counterparts, indeed becoming quite common (Fitzgerald 1990:498). Given this instance of formal, functional, and material continuity, archaeologists have been hesitant to use the presence of rolled copper-base metal beads on protohistoric sites as a diagnostic indicator of European contact. This is due to the fact that absent metallurgical analysis, it has been difficult to determine with confidence the source of the raw material.

In the early contact period, ornamental artifacts made by native hands out of European-introduced metals appear in the material assemblages of many groups who did not have long native copper working traditions (Childs 1994). Copper-base metal beads are perhaps the most widespread examples of these types of objects. Fashioned by bending (commonly referred to as "rolling") either the short ends or long edges of prepared rectangular or square blanks or strips into short cylinder or longer tube shapes, they are found on early contact sites in the Northeast in various sizes and proportions. Extremely long (60–120 mm), large diameter (5–7 mm) tubular bead forms are horizon markers for the last half of the sixteenth century in much of northeastern North America (Bradley 1987:70–71). Beads are common among the New York Seneca by 1600, and become a regular feature on Ontario Iroquoian sites by the first third of the seventeenth century (Drooker 1996b:158). Cylindrical beads appear very early in the Southeast (Drooker 1996a:Figure 3.8a; Smith 1987:37, 45) becoming a pan-regional horizon marker there from about 1525–1625.

In the Lower Great Lakes, copper-base metal beads are known from a number of protohistoric sites, including the Grimsby Cemetery (1640–1650), Ontario (Fitzgerald 1990; Kenyon 1982:xvi; Stothers 1991, 2000). Their appearance in Neutral graves at Grimsby provides several clues to their decorative and ornamental uses. In one burial, small beads appear at the knees of an adult male, indicating that he may have been wearing beaded garters or leggings at interment. Additional beads were found strung on a thin leather thong that may have supported a shell gorget located in the upper chest area (Kenyon 1982:24). In other instances, metal beads are found strung together or with single or multiple shell beads on s-twist cedar cording. They have also been recovered strung or fastened alone or with glass beads onto rawhide thongs, or massed together as grave offerings (Kenyon 1982:13, 30, 40, 90, Plates 44 and 173). Strings of beads served as necklaces and bracelets, occurring with both adults and children. Large tubular beads are recovered as well (Kenyon 1982:41).

In the Western Great Lakes, George Quimby (1966:67, 72) notes that rolled tubular beads are "typical of the entire historic period, but are most abundant in sites of the Middle Historic Period [1670–1760]." Even so, surprisingly few copper-base metal tubular beads were recovered from any period at Rock Island, Door County, Michigan, although some presumably small rolled "brass" beads are found fastened to a buckskin thong in mid-seventeenth-century deposits (Mason 1986:80, 213). Another long tubular bead was found there in later (1670–1770) contexts (Mason 1986:32).

In this same region, archaeologists have found evidence for an additional way in which small copper-base metal beads were used beyond stringing or fastening to thongs or lengths of cordage. Quimby (1966:128), Charles Cleland (1971) and Robert Hall (1993) each record instances in which pieces of leather, thought to be fragments of sashes or belts, are embellished with metal beads. In each case, small beads had been fastened to the leather in multiple rows, producing an overall mail-like effect and exhibiting various linear patterns. Use of different sizes of beads, alternate directionality of attachment, and variable row spacing are three of the ways that designs were achieved. Hall's piece, a decorated deer hide fragment, was recovered from a plowed field at the Beaumier Farm site in Brown County, Wisconsin. Delicate and now quite fragile, it holds 68 tiny beads placed tightly next to one another in four parallel, offset rows. Hall (1993:25) comments that the beads were likely attached to the material when they were in a bent "u" form and were then rolled shut in situ. All bead closures are on the same side. My own examination of it revealed that the beads are quite small, each measuring only about 2.1 mm long and about 2.7 mm in diameter. Scratch test results indicate that at least two of the beads are copper.

Similar finds come from farther into the interior (Bray 1978:58–59). Beads strung end to end and sewn on cloth were found in a grave at the late Oneota site of Gumbo Point, Missouri. The Flynn cemetery in northeastern Iowa (Oneota, 1690–1700) yielded "girdles" of juniper bark to which tiny beads had been fastened.

Finds in the Illinois Country have been far less dramatic. For the early historic period, James Brown (1990b:238) reports what he calls "brass hair tubes" from Huber (Oneota/Chiwere Siouan) contexts at the Oak Forest site, near the southern tip of Lake Michigan.

Evidence from the Haas/Hagerman site provides important insights into how the Illinois made and used these objects. Small rolled beads and tubular beads represent the largest category of finished ornamental artifacts in the industry (Figures 6.2 a, b). All but seven (*n* = 136) are small and cylindrical, and range in finished length from 1.8 mm–6.5 mm. All beads are either ovoid (*n* = 83) or round (*n* = 28) in cross section. A small number (*n* = 4) appear quite flattened. On average, small ovoid beads are slightly longer (3.7 mm) than are round beads (3.1 mm). Their diameters (ovoid = 3.7 mm; round = 2.9 mm) are slightly larger as well. The blanks used to make them are also slightly longer (ovoid blank length mean = 12.7; round blank length mean = 10.5 mm). Many blanks are not perfectly rectangular.

Tubular beads (*n* = 7) are also small. Their lengths do not exceed 27.8 mm and their diameters are no larger than 4.8 mm. A few small tubes (*n* = 3) and tubular beads (*n* = 4) of various lengths and proportions also occur (Figures 6.2 c, d). No long, large diameter tubes or tubular beads are found in this industry.

Bead closures vary. Overwhelmingly, the short ends of the blanks close by overlapping to varying degrees (*n* = 97), but some either abut (*n* = 17) or do not meet (*n* = 6) (Figure 6.2 b).

Bead fabrication is a simple procedure that can be accomplished easily by hand. All small beads are fashioned from sheet between .2–.5 mm in thickness, which has been cut into roughly rectangular or square shaped blanks ranging in length from 7.3–20.4 mm. Shear burrs are present; jagged edges are not often smoothed. Blanks are then bent and closed around a curved mandrel or form of desired diameter. Edges are pressed flush in some manner to close the bead. A few closures look as though they have been hammered down. In the vast majority of instances, beads appear to have been rolled into shape evenly and smoothly; relatively few exhibited the kinks (sharp bends) from bending stresses commonly seen in beads manufactured prehistorically (Childs 1994:238–239; Smith 1981a:78). A few, however, are quite irregular and were left twisted, looped, or bent repeatedly. Some (*n* = 25) look to have been abandoned or used in a less than completely finished condition; blanks are only partially rolled and their closures are either very open or not pressed down.

Partially fabricated beads from the Iliniwek industry demonstrate this shaping process. One incompletely rolled blank was abandoned early in the shaping of a tubular bead (Figure 6.2 e). Indentations along the body of another "finished" bead result from differential manual pressure applied to

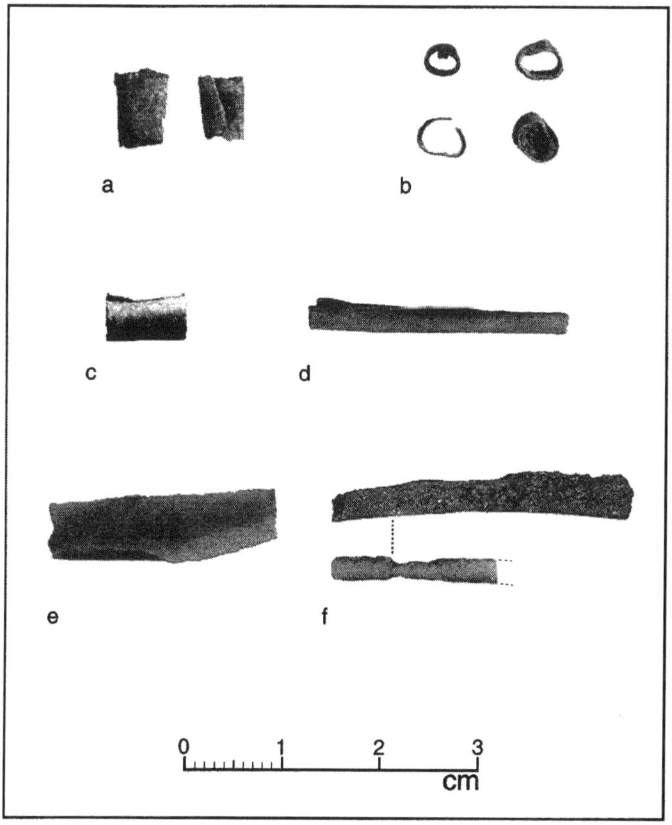

Figure 6.2 Copper-base metal beads and bead forming techniques: (a) two small rolled beads typical of the industry; (b) bead cross-sections showing closure types: top left and right, closure overlaps; bottom left, closure does not meet; bottom right, closure abuts. Bottom right bead still contains the material to which it was attached; (c) small tubular bead; (d) small flattened tube; (e) long rectangular blank being rolled into a tubular bead; (f) blank rolled into a tubular bead, probably without a mandrel; note multiple (finger?) indentations along the longitudinal axis of the piece. The deep indentation near one end of the artifact (f bottom) probably reflects the use of some type of tool to steady or compress the metal.

the piece during the forming sequence. The same bead carries an oblique indentation from what appears to be some sort of implement used to steady it, perhaps as it was being affixed to whatever it was being attached (Figure 6.2 f).

Since beads are only found in excavation levels and in refuse pits at the

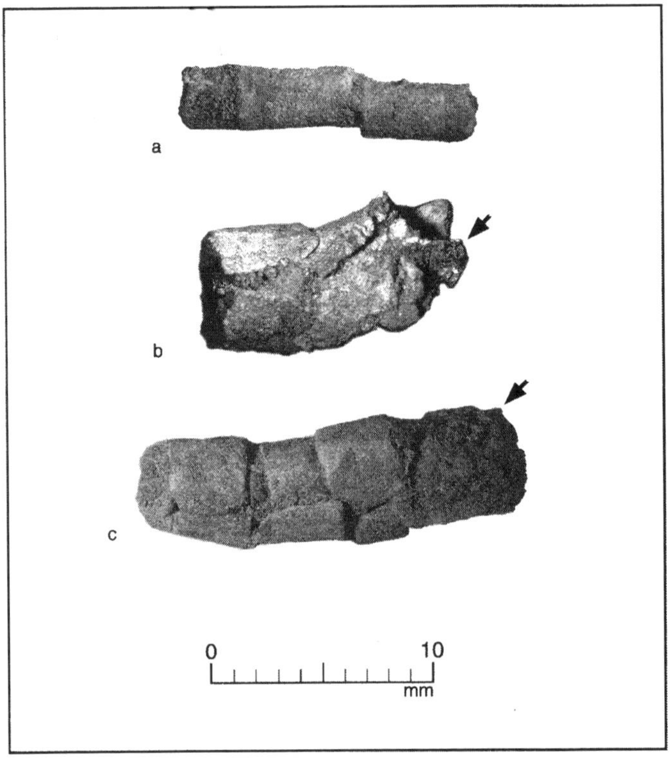

Figure 6.3 Bead use at the Iliniwek Village: (a) three small, round beads attached tightly and closely together onto organic material; (b) two larger beads encasing organic material and a tiny strip of iron (arrow), which is visible in the core of the piece; (c) three beads attached to organic material also containing a tiny strip of iron. A fourth bead has become detached.

Iliniwek site, there is little direct evidence as to how the Illinois actually used them. However, in several instances single or multiple beads were found still attached to the material around which they had been fastened (Figures 6.2 b and 6.3). The beads are affixed to thongs made of some type of unidentified organic material. A small iron wire or rod is found running along the inside of at least three of the masses (Figures 6.2 b, 6.3 b, c). While the presence of the iron rod is unusual, beads have commonly been found on sites of the period strung on or affixed to leather, rawhide strips, thongs, or twine (Bray 1978:59; M. Brown 1975:32; Hall 1993:25; Sempowski and Saunders 2001:113; Wray, Sempowski, and Saunders 1991:72).

Further information about how the Illinois used metal beads comes from the roughly contemporaneous and slightly later Zimmerman site, LaSalle County, Illinois. At Zimmerman, Margaret Brown (1975:32) also found single rows of beads strung on a rawhide strip. No small beads were found associated with the 26 individuals interred in historic period burials excavated there. However, three "brass" tubes had been placed with two children. Wray et al. (1991:248) point out that at the early contact Cameron site (Seneca) beads are almost exclusively found interred with immature individuals.

"Coiled" or "Spiral" Strip Beads

Three forms resembling "coiled" or "spiral" strip beads were identified in the sample (Figure 6.4).[3] They are similar to types found at early contact sites in Pennsylvania, New York, and Ontario (Kent 1984; Kenyon 1982:51, 105, 144; Wray et al. 1991:76, 248). These beads are made from long strips of metal wound longitudinally around some sort of dowel or leather thong. While examples from Susquehannock sites can exceed 5 cm in length, smaller specimens are found at the Grimsby Cemetery and at Seneca sites in western New York (Kenyon 1982:40, Plate 42; Sempowski and Saunders 2001:114; Wray et al. 1991:248).

The three Iliniwek specimens are small and irregular in size (two are illustrated in Figure 6.4). They measure 8.1 mm, 10.2 mm, and 12.5 mm in length, and feature 2.5, 3.5, and 5+ rotations respectively. Two of the three spirals are flattened at least partially. Like these artifacts found elsewhere, the Iliniwek examples are fashioned of long strips of metal. The prepared strip is then wound longitudinally in a single layer along the length of a mandrel of appropriate diameter. When the desired number of revolutions is reached, the mandrel can be pulled out, leaving a hollow ovoid or round coil or spiral. While many of the longer specimens from sites elsewhere are found threaded on leather thongs, no thongs occur (or survive) with the Iliniwek specimens. In one instance, however, the end of the strip itself is used as a central longitudinal axis or form for the bead, and the spiral is wound around and beyond it (Figure 6.4 b). It is visible inside the spiral.

The strips from which these objects are fashioned are manufactured by cutting a long, thin blank (the longest blank is 58.5 mm in length) off of a large piece of metal. On two of the specimens, long, shallow, nonrandom striations are clearly visible along the longitudinal margins of the blanks, possibly indicating that the blanks had been cut with a tool that left striations as

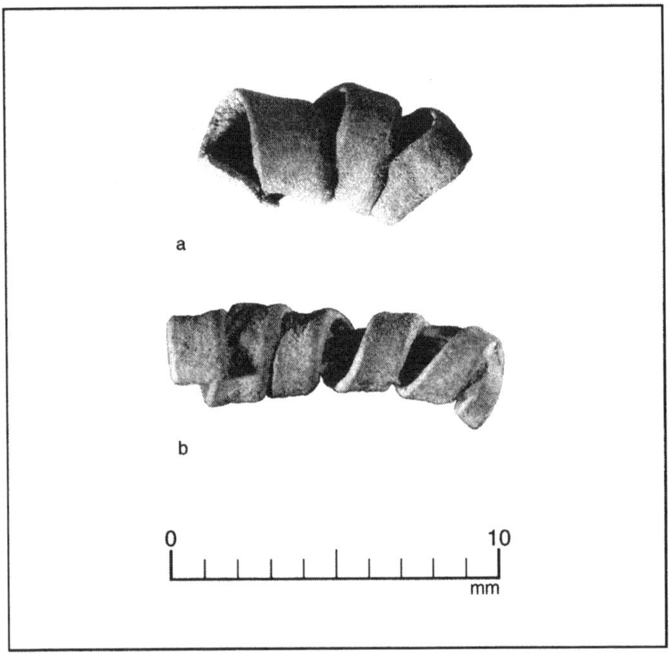

Figure 6.4 Spiral strip beads: (a) small brass bead; note fine striations along the lateral margins; (b) the blank from which this bead is made is 58.5 mm long and is coiled 5+ times around itself. The end of the strip is visible in the bead's core. Again, fine striations occur along the lateral margins of the strip.

it was pulled across the metal (Figures 6.4 a, b). Even though the surface of the piece is corroded, these striations are quite apparent to the naked eye. They strongly resemble those documented on the surfaces of drawn wire (Ogden 1991:98, 100) or those that might be found on the edges of a (machine) rolled and sheared sheet. Another less likely possibility is that they were the result of grinding burred margins smooth after shearing.

These objects appear in burial contexts among the Neutral and Seneca. In Seneca burial contexts, they are found about the head and ears, at times in vertically strung sets also containing shell beads (Karklins 1992:66; Sempowski and Saunders 2001:113–116). No direct information has as yet been found as to how the Illinois may have used these objects. However, one contemporary documentary entry may provide a clue as to their use among the living. In his description of Ottawa men, Pierre Esprit Radisson (1885 cited in Kinietz 1996:235) said,

> Their ears have ordinarily 5 holes, where one may putt the end of his finger. They use those holes in this sort: to make themselves gallant they pass through it a skrew of coper with much dexterity, and goe on the lake in that posture. When the winter comes they weare no capes because of their haire tourned up. They fill those skrews with swan's downe, & with it their ears covered.

Dablon (Thwaites 1896–1901:55:217) relates that the Illinois too wore their hair "short and erect" as did the Ottawa, but with a "long lock on the side of each ear." Whether the Illinois used the same or similar ornamental means to keep their ears covered in winter, or how relevant this practice might be either to this artifact form or to the Illinois is not known. This description might equally (or perhaps more aptly) apply to a similar form called copper coils that are described below.

Clips

These are the second most numerous finished objects found in the industry (Figure 6.5). Also crafted from flat rectangular blanks, they differ morphologically from beads. Compared to beads, which are either round in or ovoid in cross section, the midsections of these objects are decidedly flat (Figure 6.5 a). Their short ends are bent to form distinctive "legs," which appear to have been pushed through slits in a material backing then bent inward toward their midlines in staplelike fashion (Figure 6.5 b). It is likely that they were used to embellish relatively thin pieces of textile or leather rather than something with a round or ovoid core, as their finished cross sections indicate that they lay flush with rather than protruded from the surface of the material to which they were attached. There is no indication that clips were strung.

W. A. Kenyon (1982:12) recognizes similar forms attached to a decorated piece of rawhide associated with a Neutral grave at Grimsby Cemetery. Calling these forms "clips," he remarks on the manner in which blanks were folded in half and attached one at a time and apparently randomly to rawhide. "[C]opper strips were inserted half-way through slits in the hide, then bent over and squeezed flat. The result was a small copper rectangle on each side of the piece of hide."

While these objects occur in various sizes in the Southeast, in the Midwest, and on the Plains (Bleed, Toom, Johnson, Smith, and Roberts 1995:332; Drooker 1996b:160; Lindsey 1964; Smith 1987:47–48), they are much less frequently reported in the protohistoric/early historic period literature than

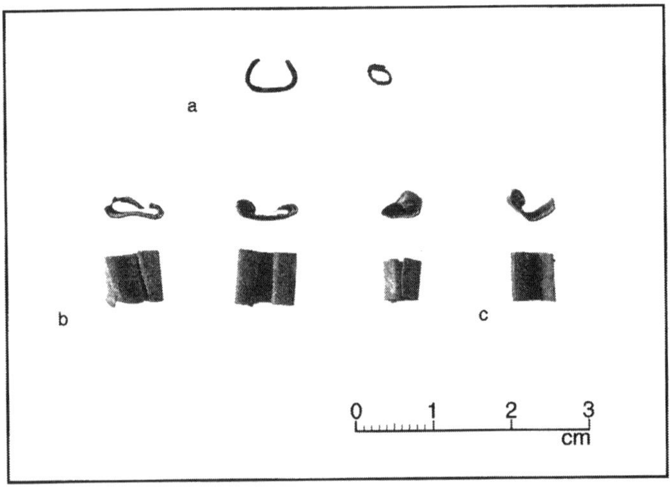

Figure 6.5 Clips: (a) comparison of clip (left) and ovoid bead (right) profiles. Left is brass; (b) examples in plan and profile views; (c) partially opened clip.

are beads. Perhaps clips are simply not used in as great abundance. It also may be, however, that although formal distinctions do indeed exist between these forms, the term "bead" may be used when "clip" is actually meant. For example, in his description of an ornamented breechclout type garment found in a Creek burial context, Ross Morrell (1965:51 cited in Walthall 1992:142) said, "The fabric was decorated with rolled brass beads, mounted 'staple-fashion' to the textile, in vertical rows one inch apart and spaced one-eighth to one-fourth inch vertically." It is unclear from this description to which formal type he actually refers. However, Peter Bleed et al. (1995:329, 332) note the distinction in form and manner of attachment when describing as clips the single row of ornaments embellishing a leather garment fragment from the protohistoric Medicine Crow site, South Dakota.

Douglas Birk and Elden Johnson (1992:223) record another use for clips among the protohistoric Dakota. Found in a burial at the Vineland Bay site, clips are utilized as decorative covers for two vertical centerline straps holding the two sides of a breast plate together. In this instance, these forms serve both functional and decorative purposes; first to hide the sewn interface between the two sides of the breastplate, and second, to enhance the visual appeal of the finished item.

The 152 clips from the Haas/Hagerman site are found in complete (n = 128) or fragmentary (*n* = 24) form. Cut and prepared rectangular blanks of roughly the same size as ovoid beads were utilized to make them. Clips are found in various stages of manufacture and use: with legs bent in, bent out (open), or partially open. Complete, closed clips averaging 5.3 mm in finished length dominate the assemblage (Figure 6.5 b). Several (*n* = 16) are irregular or occur in a twisted or bent condition. The flat, central portions of many of them undulate (Figure 6.5 b left), which gives their midsections a "rippled" appearance. This may be the result of one or a combination of the following manipulations: 1) when they were being affixed to the backing, they may have been squeezed shut so tightly that their midsections bent, or 2) in the act of removing them from use, the stress of bending (or prying) the legs back open (as we would a staple) bent the midsections. While the former explains the rippling noted on closed clips, it does not cover the fact that so many midsections of the open (or opened) clips are also rippled. This renders the latter a more feasible explanation. Either possibility explains the high number of "c," "u," "e," and "v" shaped fragments (*n* = 55) in the industry which probably represent broken clips or beads. In contrast, the bodies of beads are rarely rippled, even when closed. Clip closures either overlap or do not meet.

Twenty-seven of the Iliniwek clips appear to have been removed from a previous use. There is ample evidence from the irregular and/or partially opened positions of the legs and the bent appearance of the midsections of these artifacts to suggest that the clips had been pried open and pulled away from a prior attachment. Perhaps this was done with the intention of reusing them but the objects were discarded or lost instead.

At the same time, a number of specimens (*n* = 37) were abandoned sometime during the process of manufacture and appear to have been only partially prepared for use. Twenty-five look as though they were ready for attachment as they have both legs bent up but not in (Figure 6.5 a). An additional twelve have only one leg bent up; the other remains flat (Figure 6.5 c).

There is no direct evidence for how these clips were used by the Illinois. Clips are found in all excavated areas, both inside and outside of houses. No organic materials to which they may have been attached survive. However, their frequency at the Haas/Hagerman site indicates that they were an important means of embellishing clothing and accessories.

Tinkling Cones

Variously called jingles, dangles, bangles, tinklers, and tinkling cones, these hollow, cone-shaped decorative objects have a wide distribution in protohistoric through historic period native material repertoires. Called "noise making instruments" by Quimby (1966:73, 122), tinkling cones persist in the Western Great Lakes and Plains as decorative fringe items long after many other objects of native technology disappear (Bleed et al. 1995:327). Currently,[4] commercially available yellow and white metal conical tinklers still adorn native objects and accessories such as purses, clothing, and moccasins. Hundreds of small metal cones or large, rolled, embossed snuff can lids (also commercially available) are affixed to girls' and women's dresses made especially for powwow "jingle dances" (Hungry Wolf 1999:122–127). Originating among the Ojibwa at Mille Lacs, Minnesota, these lively yet graceful dances and the special songs that accompany them are meant to heal the body and the spirit (Dubin 1999:564).

James Bradley (1987:70, 133) claims that copper-base metal tinkling cones[5] appear among the protohistoric Onondaga Iroquois during the late sixteenth century. According to Fitzgerald (1990:503), they emerge in the Lower Great Lakes as early as 1580. He notes that among protohistoric Ontario Iroquois, these forms were meant to imitate hollowed out deer phalanx bangles common on their earlier sites. However, large tinklers of native copper were found at Late Woodland (late precontact) sites in northern Michigan and in Cook County, Illinois (Quimby 1966:39, 43). As to the appearance in the Western Great Lakes of tinkling cones made from European-derived copper-base metal, Quimby (1966:76) comments that they are one of "a number of miscellaneous trade objects that are indicative of the historic period in general and for which I cannot yet detect any specific changes from one temporal division to another within the historic period."

While there is no inclusive typology for these objects, their basic form is that of a hollow-tipped cone. They are distinguished from conical projectile points that have closed tips, and in some cases small holes near their bases, presumably to facilitate hafting (Bray 1978:37; Morse 1992:68; Quimby 1966:119). Tinkling cones are formed by rolling a flat trapezoidal or square shaped blank into a cone with an open apex. A thong or some sort of suspension material (often hair) is threaded through the open tip and looped or knotted on the inside. Sometimes tufts of animal hair attached to the thong on the inside of the cone extend out of its base to further embellish its appearance. Cones are hung (also sewn on) singly as a fringe decoration on clothing or accessories such as bags or purses. They are meant to dangle freely but to contact each other and "tinkle" with the movement of the user or the wearer. The delicate sound they make as they strike one another was undoubtedly

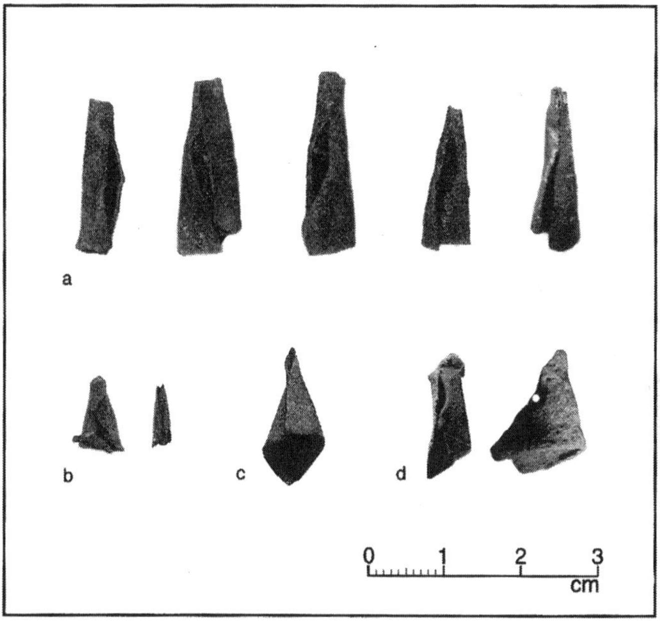

Figure 6.6 Tinkling cone forms: (a) cones with open midsections. Oblique depressions occur across the necks of all of these artifacts. Far left and second left are smelted copper; far right is brass; (b) small cones. Right is brass; (c) cone made on a square or kite-shaped blank; (d) irregularly rolled cones.

pleasing to the ear. As with beads, the shiny, reflective qualities of the metal were assuredly quite striking as well.

There are 21 tinkling cones and one tinkling cone fragment in the Haas/Hagerman sample (Figure 6.6). They vary widely in size and form. Of all the finished artifacts found in the industry, tinklers are made from the thickest metal sheet (.2–.7 mm in thickness). They fall into two distinct size groups. Five are very small, not exceeding 10 mm in length (Figure 6.6 b). The remaining cones (n = 16) range in length from 15.0–37.2 mm (Figures 6.6 a, c, d).

Aside from two finely formed examples, most of the tinklers would have to be described as crudely fabricated. This assessment is based on two major factors: 1) the irregular shape and poor condition of the sheet metal blanks selected; and 2) the rather unorthodox form that many of the finished cones actually take. At first glance, several are even difficult to recognize as cones, appearing to be the product of opportunistic rather than careful shaping technique. Uneven, bent, or irregularly shaped blanks with jagged margins are used to manufacture many of the cones, especially the small ones. In one instance, the trapezoidal "blank" used to fashion a tiny cone is so irregular and so bent that the normally open base is closed (Figure 6.6 b, left). No attempt was made to correct the situation. Three larger cones, including two made from triangular blanks and one fashioned from a rectangular blank, are quite irregularly rolled (two examples are illustrated in Figure 6.6 d). Their seams hardly close at all; two barely touch and the other only slightly overlaps. Two are flattened.

The largest group of like cones (n = 11) are made on either trapezoidal (n = 10) or square blanks (n = 1) (Figure 6.6 a). They appear to be a local production, distinctive in that they share an unusual type of closure. Unlike cones seen elsewhere whose closure seams are closed (either overlapping or abutting) from neck to base, these cones, while closed at the neck, are either open through the midsection and closed again at the base, or are open both through the midsection and the base. It looks as though the intention was to achieve an open midsection; in many cases, the midsections appear to have actually been pulled open purposefully to expose the body of the cone. Why this should have been done is not known; perhaps the tinkling sound changed when the midsections were open, or some type of distinctive decoration was placed inside the cone, which was exposed when the body of the cone was opened.

These cones share another important manufacturing signature. They all carry at least one or two oblique tool impressions across their neck closures. These indentations are concave and vary in depth. They occur on either one or both lateral sides of each of these specimens. They indicate locations where a curved tool of some kind was used to apply pressure on the already rolled piece for one of a number of reasons: 1) to hold it steady while the midsections were being pulled open; 2) to press the neck closure down more tightly and smoothly; or 3) to squeeze or press the cone more securely around the material to which it was fastened. Since at least three other cones with closed midsections also exhibit these compression signatures, it was likely done for one of the latter two reasons. In any case, this action compressed the necks of these objects slightly, giving their bodies a distinctive skirtlike appearance.

Two remaining cones produced from square or "kite-shaped" blanks are quite similar in form and resemble those previously reported by Longoria (1998:135) from Iliniwek Village surface collections (Figure 6.6 c). These tinklers have open, pointed tips and pointed bases. Like some of the examples Longoria cites, one of these two cones is flattened, perhaps purposefully.

There is no direct evidence from the Iliniwek Village for how the Illinois used these artifacts during this period. No tinkling cones were recovered from the burials Margaret Brown (1975) excavated at Zimmerman in the early 1970s. However, 13 were excavated from domestic contexts in Area A, which is dominated by historic Danner (Illinois) ware ceramics. One still had the rawhide extending from its apex (M. Brown 1975:31).

Evidence for Illinois use of these artifacts in the early eighteenth century comes from burial contexts at the River L'Abbe Mission at Monks Mound, Cahokia, Illinois (Walthall and Benchley 1987:71). This mission and its associated Cahokia Illinois settlement were occupied from 1735–1752. Nine tinkling cones were excavated there, eight of which came from burials. Considered "small" by the site researchers, these "brass" artifacts ranged in length from 9 mm to 19 mm. They were associated with a child's burial, and were located on the upper part of the right side of the body (Walthall and Benchley 1987:37–38). Several still had deer hair within them.

Triangular Perforated Pendants

These objects are flat, triangular, perforated ornaments cut from metal sheet (Figure 6.7). They fit into Drooker's "pendant" category, which includes "flat cutout shapes with a perforation suitable for hanging" (Drooker 1996b:165). All were recovered from the same pit feature in Area F, which contained one complete specimen and four fragments. Two of the fragments represent almost complete artifacts, one missing only its tip and the other its tip and one lower corner.

The pendants in the Iliniwek sample are small (11.4–18.2 mm long), quite thin, and delicate (Figure 6.7). They are clearly fashioned from blanks cut in a triangular shape, the points of which are rounded and smoothed, possibly by grinding. The perforations are equally small (between .9 and 1.5 mm in diameter) but the metal displaced by the tiny punctures was not ground or smoothed away. Unlike any of the other finished artifacts examined in this industry, the metal used to manufacture these objects appears to have been hammered down to reduce the thickness of the original sheet. In addition, the area surrounding the perforations appears to have been thinned even further than the body of the artifact, perhaps to facilitate successful puncture at such a precarious place on the artifact.

All of these artifacts are rubbed on one or both sides with red ochre, a mineral pigment long known to have been used by native peoples. Whether this is done to enhance the "redness" of these articles is not known. Redness,

Figure 6.7 Triangular perforated pendants: Four pendants, obverse and reverse views. Red ochre is found on all surfaces. The flat surfaces of many of these artifacts, especially those of the large pendant on the lower right, appear to have been hammered quite heavily. Some edges may have been ground. None of these artifacts was examined metallographically.

it will be remembered, connotes "light," "life," and the animate and emotive aspects of life (Bisson 2000:111; Hamell 1983:7; 1986:75). According to George Hamell (1986:76), it could also connote hostility and war. For the late-seventeenth/early-eighteenth-century Illinois, red meant blood, at least in the context of the disposition of prisoners. This is known through the writings of Sieur Pierre Deliette (Pease and Werner 1934:384), who said (in part),

> When they [the Illinois] want to put him [the prisoner] to death, they bring him back to the cabin of the most considerable of those who have offered him, giving the captive to them, with a kettle and a hatchet which they have colored red to represent blood.

In the context in which this relation is written, it foretells the tortures to which the person (usually a male) is apt to be put.

Whether the color red meant *only* blood among the Illinois is not known with certainty, but its meaning was probably not so narrowly confined. In his narrative of the Peouarea visit in 1673 Marquette (Thwaites 1896–1901:59:131) connects the color with war. Discussing the revered calumet, he says, "There is a Calumet for peace and one for war, which are distinguished solely by the Color of the feathers with which they are adorned; Red is a sign of war." He further notes that the Illinois painted their faces with red ochre when they prepared to go to war, and that the war "Captains" could be distinguished among the warriors because they wore red scarves skillfully fabricated of bear and buffalo hairs (these were probably dyed red) (Thwaites 1896–1901:59:127). Again according to Deliette, the faces and hair of the dead (apparently not necessarily victims of war) were also "painted" red in preparation for interment (Pease and Werner 1934:357).

Triangular copper-base metal pendants of this type do not seem to be widely known in the Western Great Lakes/Upper Mississippi Valley region during this early period. While Drooker (1996b:152, 165–171) documents several pendant forms for the early protohistoric interior, these particular forms are not among them. The closest is a long, narrow perforated trianguloid example from Madisonville with serrated edges, visually unlike these examples.

In the Illinois Country, John Walthall, F. Terry Norris, and Barbara Stafford (1992:142) report finding a single triangular pendant they call a "dangle" at the Naples site, Scott County, Illinois. The site is interpreted by these researchers (1992:149) as a terminal seventeenth-century Illinois winter hunting camp. The specimen they recovered is perforated in the center rather than at the apex, however. Cleland (1971) identifies triangular pendants with central perforations not as "dangles" but as projectile points. These items are also recognized as such by Alexandra van Dongen (1996:146), and Jefferson Chapman (1985:102) illustrates contact period Cherokee examples with remnants of hafts still remaining. Walthall, Norris, and Stafford (1992:142) argue that given the "thinness and diminutive size" of their specimen, it would not be suitable as a projectile point. Alternatively, they suggest that it was likely sewn onto clothing as a decoration.

The Iliniwek Village triangular pendants are most congruent with potentially later examples from the northern Great Lakes and from the Utz site (Missouri/Oneota) in Missouri. Ronald Mason (1986:31, 37) recovered two artifacts of like form at Rock Island, a multicomponent protohistoric/historic period site at the mouth of Green Bay. They are associated with a 1670–1730 historic Potawatomi occupation there (Mason 1986:217). Bray (1978:53) reports a number of somewhat similar artifacts from the Utz site but they do not appear to have been as well made as the Iliniwek examples.

Notable also are these objects' resemblance in shape and perforation placement to melted glass ornaments reported by Margaret Brown (1972) for eighteenth-century sites in the Upper Missouri region, the Illinois Country, and Wisconsin. One such object was recovered from a burial in what was then termed an early historic "Swanson component" at the Hotel Plaza site in the upper Illinois River valley near Starved Rock (Schnell 1974). Gail Schnell (1974:59) and Margaret Brown (1972:433) both suggested that this context might be associated with the Illinois. Schnell (1974:59), however, hypothesized a pre-1680 date for the component while Margaret Brown (1972:433) stated simply that it was "pre-1732."

Again, no direct contextual assignments can be made concerning how these artifacts were actually used by the Illinois. However, there are several indications that they may have indeed been used as dangling or sewn-on decorations or were at least being prepared for that use. First, the perforations on four of the five examples are broken (see Figure 6.7). Upon close examination, they do not appear to have been broken in manufacture. Thus, pendants may have broken or fallen off while in use, having been sewn onto or hung from clothing or attached to the body, perhaps to the hair or ears. Second, ochre had already been applied to the surfaces of the artifacts, not surprising if individuals were enhancing their appearance in war or for mortuary ritual.

Bracelets, Rings, and Coils

Made from metal strips, tubing, or wire, bracelets and rings are found quite frequently in early contact period assemblages in the native Northeast and Western Great Lakes/Upper Mississippi Valley. In the Lower Great Lakes bracelets, rings, and conical bangles appear very early (1580–1600); bracelets and rings are most prevalent in pre-epidemic times (1580–1632) (Fitzgerald 1990:559). David Stothers (2000:54–55) cites several metal rings from the Indian Hills site, near the southwestern end of Lake Erie, dating between 1550–1643. According to Fitzgerald (1990:497) the appearance of all of these types of items is evidence of the elaboration in personal ornamentation that occurred as the commercial fur trade became more firmly established. Most early examples are recovered from burial contexts.

While bracelets and rings were not among the earliest forms of reworked copper-base metal to appear on Onondaga Iroquoian sites, they join the material repertoire there in the first half of the seventeenth century (Bradley 1987:133). During that same time period, they are also found at the Ossossane (Huron) ossuary where Father Jean de Brébeuf witnessed a "Feast of the Dead" in 1636 (Kidd 1953:369–370). Jesuit Paul Le Jeune would later say of their use among the Huron, "In France, bracelets are worn on the wrist; but the Savages wear them not only there, but also above the elbow and even on the legs above the ankle" (Thwaites 1896–1901:44:287–291).

By the 1630s, copper-base metal bracelets were also popular in Seneca mortuary contexts; multiple examples adorn the arms of interred individuals (Wray et al. 1991:75). Rings appeared in the hair, around necks, and also on fingers (Wray et al. 1987:51, 59, 141). While necklaces containing copper-base metal beads are frequent at the Grimsby Cemetery (Neutral), copper-base bracelets and rings are not. These items occur much more frequently in other media such as shell beads, cedar bark, and iron than they do in tubing or solid copper wire (Kenyon 1982:55, 61, 63, 95, 138, 169, 184, 187, 195, 198).

Most of the examples in the Western Great Lakes come from a somewhat later period. Simple C-shaped bracelets are diagnostic of the Early and Middle Historic periods in the region (Quimby 1966:72). They are found in both mortuary and nonmortuary contexts and appear to be of the same type as those documented farther to the east (Quimby 1966:122, 127). Several wire bracelets were recovered from the earliest occupations (1641–50/51) at Rock Island (Mason 1986:44). Some bracelets, however, are complex and quite elaborate (Cleland 1971:28). One from the Lasanen site was formed by twisting two strands of 1.5 mm solid copper-base metal wire tightly around each other into one long, single twisted strand. This strand was then doubled and bent into a C-shape.

Most copper-base metal bracelets are much simpler variants of this C-style, which Fitzgerald (1990:505) appropriately characterizes as "rudimentary" in construction. Strips or bands of cut copper or brass sheet of various widths are simply formed into an open bangle or C-shape. Some have overlapping ends and are incised (Wray et al. 1991:75, 249). Bracelets and rings fashioned of solid wire or tubing are often made by snipping off a small length of material and bending it into the desired size and shape.

The Iliniwek sample contains five complete rings and one C-shaped bracelet fragment (Figure 6.8). They are all made of hollow tubing. Four of the rings and the bracelet are fashioned of a distinctive double-lobed tubing commonly referred to as "B-wire" or "butt convoluted wire" in the midcontinent because of its characteristic "B"-shaped cross section (Birk and Johnson 1992:223; Bray 1978:31; J. Brown 1961:61–62; M. Brown 1975:32; Drooker 1996b; Wedel 1959:63–64) (Figures 6.8 a, c, d, e; see also Figure 7.2 b). Technically, this is a misnomer. If, by definition, wire is "solid metal of uniform cross section," this material should not be called wire at all. Hollow cylinders,

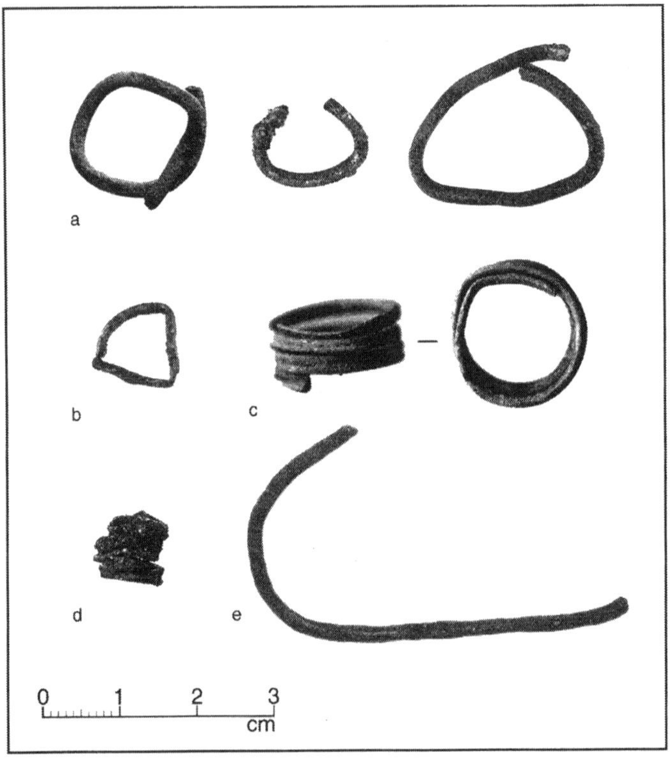

Figure 6.8 Rings, coils, bracelet fragment: (a) rings made of double-lobed tubing; (b) ring fashioned from single-lobed tubing (smelted copper); (c) double-lobed tubing coil in plain and profile views; (d) double-lobed tubing coil in plan view; (e) fragment of a c-shaped bracelet made of double-lobed tubing.

no matter how small in diameter, are more appropriately referred to as tubes (Untracht 1968:42).

The remaining ring is shaped very crudely from a length of single-lobed tubing (Figure 6.8 b). This type of tubing is also called "e-" or "o-wire," because of its shape in cross section. The finished diameter of both types of tubing ranges from 1.6 mm to no greater than 2.7 mm.

Fitzgerald (1990:507–508) has identified a number of types of tubing used in bracelet and ring manufacture in the Lower Great Lakes, and the "B"-shaped, or butt-convoluted type, and the "o" and "e" types are among them. Ornaments formed from these materials are found in the earliest period of European material influence in the region (1585–1600). A coiled bracelet of double-lobed tubing occurs at the Factory Hollow site, a 1610–1625 Seneca burial and village locale (Sempowski and Saunders 2001:396). One distinctive C-shaped bracelet was found at the Grimsby Cemetery and Fitzgerald cites additional ring finds in the Lower Great Lakes (Fitzgerald 1990:508; Kenyon 1982:198, 215).

Farther west, a circular bangle type bracelet of large diameter (3.1/5.5 mm) double-lobed tubing is reported from the Western Great Lakes at the Vineland Bay site, Minnesota (Birk and Johnson 1992:222–223). This bracelet was found near the humerus of the interred individual.

The rings in the Iliniwek sample vary in size. One appears too large to be a finger ring (Figure 6.8 a, right). It is flat on one side, either from manufacture or use. Its edges are cut cleanly and overlap. There is no evidence of the ends being compressed or squeezed as the tubing was cut. In contrast, the ring fashioned from "e"-shaped tubing (Figure 6.8 b) seems to fit only a child. It has two major kinks, giving it almost a triangular look. It is very corroded.

The rings and bracelet in the Iliniwek sample look to have been formed in the same simple manner as described above; a desired length of tubing is simply cut off a larger piece and bent into the intended shape. Overall, these specimens look to have been easily and not very carefully shaped. There is no reason to suspect that they are not made locally.

There are four coils in the sample (Figures 6.8 c, d). These artifacts are also made of double-lobed tubing and are manufactured in the same fundamental manner as the rings and bracelet. In the case of at least two of the coils, a mandrel or dowel is likely utilized to maintain evenness in the bending process. Objects are shaped by winding a length of tubing around and around a round cylindrical mandrel until the desired length is achieved. The result looks like a modern bedspring. Unlike some of the rings, which are angular and crudely circular, these coiled artifacts were, for the most part, smoothly and evenly wound into a helical form. Only one has uneven revolutions, the end product looking more like a lopsided, stylized beehive rather than a bedspring (Figure 6.8 d).

Of all the finished ornament types in the sample, we know most about the contexts of use of these tubular coils among the Illinois during this period. Artifacts of this design are well known in the early contact native Northeast and midcontinent (Drooker 1996b:161). One way these forms may have been used among the living has been discussed above. James Brown (1961:62) reiterates this possibility citing an example from the eighteenth-century Natchez, who according to Le Page DuPratz, used them as ear ornaments. In mortuary contexts, they are frequently found near the ears or in the hair. Long coils wound around wooden dowels have even been termed possible "hair pullers" or "tweezers" (Kent 1984:209; Mason 1986:37). Another unique, composite form is known as a "hair band" (Fitzgerald 1990:515–517; Kenyon 1982). Based on Illinois mortuary data, Margaret Brown (1975) has suggested that coils were used by the Illinois as hair ornaments or as headdress decorations. At the Zimmerman site, coils have been found associated with crania (M. Brown 1975:32). She (1975:39, 41) reports that multiple brass coils were found associated with four historic period burials excavated there. Three of the four burials were of children. The remaining burial was of an adult male.

Other Finished, Unmodified Artifacts

The sample also includes one copper-base metal sheet bell fragment (not illustrated) and one awl (see Figure 7.7 a). The bell fragment is the lower, bowl-like hemisphere of a hawk bell. No part of the equatorial seam is present. Although dented, the fragment maintains its overall shape. Its surface is undecorated. There are two large holes in the base of the hemisphere connected by a wide slit. The slit and holes may have been enlarged or modified in some manner but it is not possible to say with certainty.

Bells are thought to have been attached to leggings and other articles of clothing or worn around the ankles. They were a widely circulated trade item throughout the contact period (I. Brown 1979:201). The distinctive "Clarksdale" type bell appears in the interior early on at Madisonville, but this is the most northerly point at which they are found (Drooker 1996b:150). Other types of bells are, however, found after 1670 much farther to the north at Hotel Plaza (Schnell 1974:42–43) and Lasanen (Cleland 1971:27). Cleland divides the bells at Lasanen into small (12.5 mm in diameter) and large (17 mm in diameter). The Iliniwek bell is 12.7 mm in diameter.

The awl's form is easily recognizable as a piercing tool found in native tool kits since Old Copper times (see Figure 7.7 a) (Martin 1999:230–231; Vernon 1990). It has an ovoid body whose point narrows at the neck to a fine point. It measures 26.3 mm in length, 4.0 mm wide, and 2.3 mm thick. While it may have been curated from an earlier period, it is not likely.

The awl is of particular technological interest. Beyond the fact that it is solid and does not appear to be made of smooth, cut sheet as are the remainder of the items in the industry, there is one important visual clue that points to the possibility that it was made of native copper. It carries an important signature of what Ursula Franklin et al. (1981:26) have called "folding." Folding is a major step in native copper artifact manufacture that involves turning a hammered piece of copper over on itself and continuing to hammer (potentially repeatedly) until the desired thickness and dimensions are reached. This leaves a "seam"[6] or "lap" in the metal often visible on the surface of the artifact. Usually, annealing accompanies this operation. A seam was present on this artifact.

One other artifact that deserves attention here is a small bipointed object measuring only 8.9 mm in length by 2.0 mm in diameter (see Figure 7.8 a). It is clearly a finished artifact, but its function is unknown. It too looks to be made of native copper.

UNFINISHED/PARTIALLY PROCESSED ARTIFACTS

Blanks

Blanks are defined here as pieces of cut metal sheet whose form and worked condition indicate that they were intended for and/or were in some stage of being formed into finished artifacts when they were abandoned. As such, they represent an intermediate stage in the manufacturing sequence (Franklin et al. 1981:33). A total of 279 flat and manipulated blanks, including 108 blank fragments (Figure 6.9) have been identified in this sample. In some cases, once-finished artifacts may have been bent back into blank form.

Also in this category are 108 fragments of items that could not be identified with certainty as to type or to finished/unfinished condition. Seventy-seven, or 72 percent are short ends of small rectangular blanks that are rolled, bent, or folded and rolled. These are most probably parts of broken clips or beads. They may also represent manufacturing or use breakages. The remaining 31 artifacts are fragments of flat rectangular blanks.

Complete blanks come in a variety of shapes, including rectangular (n = 162), trapezoidal (n = 6), square (n = 2), and parallelogram-shaped forms (n = 1) (Figure 6.9). Their shapes are rarely regular. They are either flat or in a number of bent conditions. As such, they provide ample opportunity to observe technical processes of artifact manufacture.

Blanks are cut out of larger pieces of metal. The margins of several large blanks are uneven, suggesting that smaller blanks had been cut from them. The outline of removed material is often quite apparent (Figures 6.9 b, c). Many of the blanks, both large and small, have various size "nicks" into their margins where the ends of shearing tools have gone too far and snipped into one piece of metal while cutting another off (Figures 6.9 b, right, and c, left).

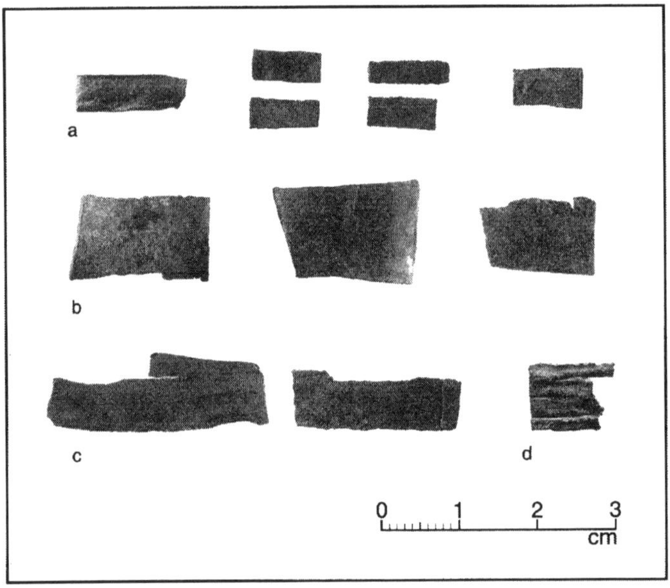

Figure 6.9 Blanks and blank removal: (a) small rectangular blanks from which clips and beads are made. Deep shear burrs occur on their edges; (b) trapezoidal blanks. Signatures of individual cutting actions are evident on these artifacts. Center and right are both smelted copper; (c) large, rectangular blanks from which blanks were removed. Left was likely abandoned during blank cutting operations (smelted copper). One lateral edge is also "squeezed." The short end of the center example is folded over. While folded edges are not atypical, the reason is unknown; (d) the only clear example of a blank that has been scored in preparation for removing smaller blanks or strips (smelted copper). The left edge of this blank is also folded.

Successive shearing actions can be seen along the edges of some long pieces (Figure 6.9 c, right).

Sheets were most likely sheared with tin snips or scissors or by the downward impact of a metal chisel or knife blade. Scoring and snapping the material apart was probably not a frequently used shearing method since deep score marks were only detected on one piece of material in the entire industry (Figure 6.9 d). The possibility that the material was simply bent back and forth until it fractured is even less likely, since the edges do not have the bent and "rippled" effect this action would likely leave on thin metal. In contrast, characteristic shear burrs were routinely visible on one or several edges of all of the objects, indicating that the sheet was cut with some sort of sharpened object. In rare cases, under the proper raking light, the actual blade channel can be seen along the blank margins (Figure 6.9 a). In some instances, the sequence in which the four sides of small rectangular blanks were sheared is visible by the way the burr pattern overlaps. From time to time, the sharp edges of the shear burrs were hammered down or ground smooth.

The notion that these artifacts were sheared with tin snips, scissors, or some type of sharpened edge could lead to the assumption that they must have been produced by Europeans because native peoples did not have access to scissors or snips. This is not the case. These objects are found at several locations, including the Ossossane ossuary (Kidd 1953:128). In the Western Great Lakes, scissor parts have been recovered from Lasanen (Cleland 1971:20, 22) and Rock Island (Mason 1986:42). Blades were, however, more generally available on clasp, sheath, or utility knives, which were in much wider circulation (see also Cleland 1971:19–21). While no recognizable scissors or scissor parts have been recovered to date at the Haas/Hagerman site, there are many examples of iron knives and knife blade fragments that may well have been used.

Shearing with a sharpened cutting edge does not account for all of the margins, however. Some are quite jagged and there are no visual indications as to how some pieces were sheared. "Tearing" does not appear to be an adequate solution (Bray 1978:31). The option remains open that some shearing could have taken place with stone tools. Mason (1986:88) has suggested that burins like the ones he recovered at Rock Island may have been applied to that purpose.

Large and small rectangular blanks account for 95 percent of the complete blanks in the industry (Figures 6.9 a and b). This is not surprising since most of the artifacts in the industry, including beads, clips, and other items (including tubing described below) are made from strip and rectangular blanks. Each of the 12 large blanks is certainly of sufficient size to yield at least four, perhaps eight smaller rectangular blanks (Figure 6.9 b). One large blank is actually grooved longitudinally in preparation for producing narrow strip blanks (Figure 6.9 d). No blank found in this sample is long enough to fashion the tubing in the lengths in which it occurs in this industry.

The small rectangular blanks in the sample are clearly clip or bead preforms. Blanks of this type are quite numerous; they are found in both complete (n = 150) and fragmentary (n = 108) form. A significant number are cut quite irregularly. This is not at all unusual, as finished clips and beads are often found to have been made from blanks of uneven width. Most of the blanks in the sample are flat (n = 119), having been discarded in an unmodified preform state. However, some have been further manipulated in some manner (n = 31). Working may be as simple as bending slightly, but more frequently they have been rolled from one short end, twisted, folded flush, or looped in an irregular manner. Whether these actions are directed toward creating an actual finished product is not known; certainly, it would have been a simple matter for the metalworker to straighten them out and begin again. The surface topography of some of these blanks is quite undulating and uneven, indicating that some degree of manipulation had deformed them somewhat before they were cast aside.

Some of the flat trapezoidal blanks may have been intended for tinkling cone manufacture. They fall within the rather wide range of blank sizes and shapes used for tinkling cones formed locally at Iliniwek. However, it is not possible to make a confident attribution.

Tubing and Solid Wire

Two varieties or "configurations" (to use Fitzgerald's term [1990:507]) of hollow tubing, double, or B-shaped, and "e" or "o" shaped tubing, appear in the Iliniwek metalworking industry. Most pieces are in fairly good postdepositional condition, but some are heavily corroded or charred.

Thirty-four segments of double-lobed tubing (totaling almost 80 cm of material) were found in the sample (see Figure 7.4 a). The tubing is made from strips of metal about 10 mm in width and .2–.3 mm in thickness. As explained earlier, the lateral margins of a strip of desired length are rolled into the midline, creating a tube with a hollow, B-shaped cross section. The finished widths of this tubing vary considerably from sample to sample, and range from 1.2–2.6 mm. In addition, the lobes are not always of equal size. The delicacy of this tubing and the purported uniformity with which it is formed has led several researchers to question its source as a native production or as a European import (Bray 1978:33–34; Drooker 1996b:161).

Some researchers (including this author) have not been entirely convinced of its European origins. Fitzgerald (1990:507–509) claims to have attested at one time to its foreign source, then reversed himself. After further investigation into early contact period native copper-base metal artifacts, he now argues that "the progression evident in the elaboration of sheet metalworking suggests that it [the intricacy of more ornate forms] was a development in native skills that reached a zenith for bracelets by GBP2 [Glass Bead Period 2, 1600–1632]." Even so, the tightness and evenness of fabrication along the entire length of the segments raise doubt as to whether achieving a finished product of such fine, albeit irregular and inconsistent width (1.2–2.6 mm), could be achieved without the use of some sort of a drawing technique. While reported by Frank Cushing (1894) for the historic Zuni, use of drawing by prehistoric native North American metalworkers has never been confirmed (Martin 1999). Close microscopic examination of the Iliniwek samples revealed no telltale longitudinal surface striations that would have provided important evidence for drawing.

Additionally, there are 11 segments of "e"- or "o"-shaped single-lobed tubing, or 154.3 linear cm of this material, in the industry. Segments vary in length. Frequently, close visual or even microscopic inspection of the tubing is required to tell the two subtypes apart. The most elaborate artifacts made from this tubing are coiled "hair bands" found associated with the cranial area of burials in the Lower Great Lakes (Fitzgerald 1990:516). These are thought to be European-made. Fitzgerald (1990:517) suggests that they could have been disassembled and the coil cut up and used for other purposes.

While the origin of the tubing remains unknown at this juncture, it is possible to make some statements about how it was manipulated at the Iliniwek Village. First, the manner in which segments are cut off longer pieces varies considerably. Most of it is simply "chopped" (to use Jeffrey Brain's term [1979:193]) off of larger pieces. Opposite ends of the same piece may be sheared differently. Some edges are cut cleanly, revealing clearly the tubing's cross section. In other instances, seemingly dulled instruments are dragged across the specimen while pressure is being applied. This leaves the edges jagged and distorted. The ends of additional specimens are compressed as a result of shearing. In multiple cases, depressions are left by some sort of curved tool that had been used to hold the object down or in place while it was being cut.

In addition to shearing, these materials seem to have undergone no little degree of manipulation before abandonment. Most pieces are at least curved; in at least two instances, segments are actually bent or looped in teardrop fashion. Many segments are twisted and misshapen, some to a sufficient degree that the double lobes are pulled apart. Kinks (yield points) are frequent. Stress corrosion cracking is noted at the kinks and at some cut ends. This condition can be quite severe.

Lastly, but importantly, one segment of double-lobed tubing was found twisted and hammered completely flat (see Figure 7.4 b). Why this was done is unknown, as there are no finished artifacts in the sample, or in the industry for that matter, made of flattened tubing. Margaret Brown (1975:32–33) reports a "brass snake" from the Zimmerman site (Kaskaskia subgroup of the Illinois), which was made from a segment of flattened double-lobed tubing and another in the process of manufacture. This object form is not uncommon in protohistoric and contact period times in the midcontinent (Drooker 1996b:161; Fox 1991:6). The Great or Horned Serpent's symbolic links with hunting success, good/evil, hurting/healing, and with copper have been discussed (chapter 4, this volume; Fox 1991:5, 1992:29–30; Martin 1999:200–204). Margaret Brown (1975:30, 32–33) suggests that snake effigies may have had symbolic and ceremonial meaning for the Illinois, since late-seventeenth-/early-eighteenth-century European visitor Sieur Pierre Deliette had reported that some Illinois regarded them as dangerous manitous. According to Deliette, "medicine men" in the village kept defanged rattlesnakes and used them in their healing "juggleries" to evoke fear in young believers and to "hoodwink" them into obedience (Pease and Werner 1934:369–375). These segments are clearly not serpents "in progress."

Unless hollow tubing goes unrecognized in archaeological reports, solid copper-base metal wire appears much more frequently on early contact sites than does the tubing of either variety. It is unusual then, that only two short pieces of solid wire were found in the sample. They are both found in the same pit feature outside House 2. They measure 4.3 and 5.8 mm in length, and are .8 and .9 mm in diameter, respectively. These segments may mend, but corrosion precluded a definitive resolution.

Scrap/Wastage

Scrap is defined here as various shaped segments or pieces of rejected copper-base metal material, also termed "debitage" or "wastage." The sample of scrap from the Iliniwek Village comes in a wide variety of shapes, sizes, and thicknesses (Figure 6.10). Scrap is classified as irregular (n = 62), irregular with two parallel sides (n = 11) or one right angle (n = 6), triangular (n = 20), rectangular (n = 17), trapezoidal (n = 15), or parallelogram-shaped (n = 3). It is important to point out here that the "shape" of the scrap refers to the form to which it might only *roughly* conform. Special attention was paid to the occurrence of two straight (cut) parallel margins or right angles on these materials. Parallel edges may indicate that blanks were being cut from these pieces. Specimens with one right angle were also placed in their own category because these

Figure 6.10 Scrap: (a) irregular scrap, folded, rolled, perforated (brass); (b) thick, irregular, twisted scrap; (c) flat, triangular scrap; (d) flat, trapezoidal scrap, ochre present; (e) flat, trapezoidal scrap (smelted copper); (f) irregularly shaped, perforated scrap, some edges irregularly folded (smelted copper); (g) large, irregularly shaped scrap, bent, folded, rolled. Individual shearing actions apparent on lower edge (brass).

artifacts could well represent the unused portion of a larger, irregularly shaped sheet from which blanks were cut. Kite-shaped (n = 1), square (n = 2), even roughly ovoid (n = 1) forms also occurred. In some instances, it was difficult to distinguish "blanks" from "scrap"; however, no recognizable finished objects, either complete or fragmentary, were placed in this category. All pieces of scrap are considered "complete" rather than "fragmentary."

Overall, scrap is the largest sized material in the sample next to tinkling cones, which are made from relatively large sized blanks. The thickest piece of material (1.1 mm) in the entire sample is also in the scrap category. At the same time, no piece of scrap is larger than 65.7 mm in length by 29.1 mm in width. Even so, the mean length of all the scrap is only 12.4 mm; the mean width is 7.3 mm. This evidence alone is not adequate to assert that material was coming to Iliniwek in small pieces. It may be that all larger pieces were cut up soon after they arrived there. The largest piece of scrap found on the site to date (from House 5) measures 115.3 mm long by 21.0 mm wide, and ranges in thickness from .4–.7 mm. Its relative thickness indicates that it is not the same stock as that which comprises the bulk of the Iliniwek sample.

While some scrap in all categories appears hardly worked beyond shearing, many pieces were manipulated in multiple ways before discard. Material is found sheared, folded, perforated, rolled, hammered, and/or bent repeatedly (see particularly Figures 6.10 a, f, and g, and Figures 7.1 and 7.3). One piece of scrap was even found folded and crushed into what could best be described as a ball or a lump. Margins are cut or sheared carelessly, folded over (at times quite flush), hammered down, bent, snipped into or "nicked," at times quite deeply. Sequential shearing actions can sometimes be seen along the margins. Some are heavily manipulated to no apparent intended end. A few pieces of scrap even appear to have even been rubbed with ochre (Figure 6.10 d).

The surface texture of scrap material is frequently uneven. There are ample and varied reasons for this, including repeated bending-crumpling/reflattening processes, intense manipulation, or simply corrosion. The metal itself is sometimes found in poor and differentially corroded condition. Corrosion is preferential in areas of most significant manipulation.

The sample was also examined to determine whether any of the scrap was actually used for some type of expedient purpose. Recent examination and reevaluation of "scrap" from Huron sites (Latta, Thibaudeau, and Anselmi 1998) have demonstrated that not all of it was rejected material, rather, some 50 percent of it was used as expedient tools. Tools were created by thinning, blunting, folding, and rolling the portion of the overall piece to be used while ignoring the rest. Tools were recognized as such by shape and the presence of use wear along the margins. Martha Latta and colleagues (1998:179) conclude that "much of this irregular metal was not 'scrap'—i.e., wastage—but rather . . . it was a bank of raw material which could be adapted for a variety of purposes at need." Microscopic examination of the Iliniwek scrap revealed no examples of such use. This may be due to the overall diminutive dimensions of the Iliniwek metal, which is generally much thinner and smaller in size than the Huron materials. While this decreases the suitability of the Iliniwek materials for such purposes, it certainly does not preclude it.

Copper alloy scrap is commonly the first European derived material to appear on sites in the Lower Great Lakes region. According to Fitzgerald (1990:515), unlike glass beads, which were consumed before going inland, scrap also found its way into the interior early on. Likewise, in the Illinois Country, it appears at New Lenox in northeastern Illinois by the 1620s (Lurie 1998). The scrap from New Lenox looks no different in form or condition from the material found at the Iliniwek Village.

OBSERVATIONS ON "STYLE" UNFOLDING

In the narrative of his stay at Peouarea, Father Marquette (Thwaites 1896–1901:59:123) had the following to say about the native-made apparel and items of personal adornment he encountered there:

> After [the] feast, we had to go to visit the whole village. . . . While we walked through the Streets, an orator Continually harangued to oblige all the people to come to see us without Annoying us. Everywhere we were presented with Belts, garters, and other articles made of the hair of bears and cattle, dyed red, Yellow, and gray. These are all the rarities they possess. As they are of no great Value, we did not burden ourselves with Them.

Marquette and his party were clearly not impressed by the "value" of the native-manufactured items showered upon them. However, in revealing copper-base metal use, we also shed important light on the kinds of "rarities" Marquette dismissed. When viewed from indigenous, rather than Western, perspectives of worth and meaning, technical knowledge, skill, execution, and social context of use, we begin to understand the meanings these objects held in native life. Regrettably, we are left with only tantalizing hints of what these artifacts may have actually looked like or how they might have been embellished with copper-base metal. The dearth of available archaeological and documentary evidence precludes our ability to reconstruct and interpret them fully, but as this chapter has shown, technical and contextual analysis of the industry from the Iliniwek Village sheds considerable light on the problem (see Paterek 1994:xii). Decorative items appear to have been crafted for the purpose of embellishing apparel, accessories, regalia, or the personal self, most probably including the very kinds of "rarities" of which Marquette speaks. It also demonstrates clearly that the Illinois themselves were manufacturing these ornamental artifacts at the site, and reveals something of how they were converted from sheet metal to ornamental objects. In this way, Illinois metalworking "style" begins to emerge.

7 / Finding "Style" Beneath the Surface

Artifact Composition and Manufacturing History

BRINGING MATERIAL PREFERENCES AND TECHNOLOGICAL PROCESSES TO LIGHT

We now turn to asking crucial questions about material selection, preferences, distribution, and working techniques that cannot be answered with confidence from formal and contextual analysis alone. Compositional and metallographic inquiries are aimed at 1) characterizing reliably the compositions and the sources of the materials that the Illinois selected and used, and 2) defining more precisely the range of production processes they employed to transform metal sheet into finished products. Using these methods, correspondences among artifact types, raw material makeup, and forms and sequences of manipulation are revealed. In discovering and assessing these links, important information comes to light concerning the kinds of technological choices and activities that served to inform, initiate, or resist processes of transformation during this period (after Childs 1994).

This chapter describes the 75-artifact data set utilized in this phase of the investigation. The results of the PIXE compositional analysis are presented first, since knowing confidently the elemental content of the copper-base metals informs the metallographic interpretations. The metallographic and INAA results follow. Lastly, the results of the metric, microscopic, compositional, and microstructural analyses are pulled together in a synthesis of what has been learned from these investigations about what the Illinois were making with copper-base metals, how and in what sequence these objects were fashioned, and what, if any, their preferences were with regard to specific metals.

SUBSAMPLE COMPOSITION

The 75-artifact subsample selected for further technological and compositional examination by PIXE, INAA, and metallography (see Ehrhardt 2002: Appendix C) was chosen judgmentally from within the overall sample. It is made up of examples from most major types of finished and unfinished artifacts. Finished artifacts (n = 26) include clips (n = 11), beads (n = 7), tinkling cones (n = 4), a spiral strip bead (n = 1), a ring of single-lobed tubing (n = 1), an awl (n = 1), and the small bipointed piece (n = 1). No examples of triangular perforated pendants or hollow tubing coils are included in the sample. Objects in the unfinished artifact category (n = 49) include scrap (n = 21), various shaped blanks (n = 19), and single-lobed (n = 4) and double-lobed (n = 5) tubing.

These particular specimens were chosen from the overall sample because they bore unique or representative characteristics of particular technological interest. For example, one of the particular tinkling cones was selected so that the deformations at locations where oblique impressions crossed their neck closures could be investigated metallographically. A clip that had been shaped by what looked like folding a trapezoidal blank in half, hammering it flush, then rolling it into final shape was also chosen to investigate that process more fully. The small awl and the bipointed piece were selected because they were suspected of being made of native copper.

RESULTS OF THE PIXE COMPOSITIONAL ANALYSIS

As related in chapter 3, PIXE is one of a number of elemental analysis techniques that have demonstrated success in distinguishing among native copper, European-derived (smelted) copper, and brass (Ehrhardt, Nash, and Swann 2000; Fleming and Swann 2000; Gersch, Robertson, Henderson, Pollack, and Munson 1998).[1] Native copper is known to be very pure (usually over 99.3–99.9 percent), containing very low levels of trace elements. Commonly, these include silver (Ag), arsenic (As), iron (Fe), and mercury (Hg) (Hancock et al. 1991; Wayman 1989:4; Wayman and Duke 1999:55; Wayman, Smith, Hickey, and Duke 1985:367). While researchers (Maddin, Wheeler, and Muhly 1980: 223) had argued at one time that this characteristic is insufficient to distinguish native copper from "fairly pure" smelted copper, investigators working on North American copper-base metal have since gained considerable ground on the problem. They have concluded that with identification and consideration of the proper diagnostic elements, native copper can indeed be discriminated compositionally from copper in its smelted form (Fitzgerald and Ramsden 1988; Hancock et al. 1991; Wayman et al. 1985). Wayman and colleagues (Wayman et al. 1985; Wayman and Duke 1999) have emphasized, however, that a two-pronged approach that combines structural evidence obtained through metallography with compositional information is a preferred strategy.

As a major binary alloy of copper, brass contains varying amounts of zinc as the predominant alloying element. Other impurities in much lesser quantities (even trace amounts) may also occur in the finished metal by accidental or purposeful addition, rendering it compositionally inhomogeneous (Pollard and Heron 1996:196, 211). While elemental levels of zinc are generally not helpful in distinguishing native from smelted copper (Hancock et al. 1991:80), zinc added to copper in sufficient amounts (usually over 1 percent) signals purposeful alloying. In this work, absolute zinc levels above the highest detection limit of ≤ 1.86 percent (as measured by PIXE) were taken to indicate brass. In this study, PIXE did not detect zinc levels less than 1.86 percent Zn. This is well above its typical detection limits of < .57 percent (Fleming and Swann 2000).

Of the 11 elements measured using PIXE spectrometry, eight, including copper, zinc, tin, arsenic, lead, silver, nickel, and antimony, are of diagnostic interest here. Iron, sulfur, and chlorine are likely constituents in corrosion products and are less useful. The results of this analysis are summarized below.

The Material Character of the Artifacts

Native Copper

PIXE results indicated that 41 of the 75 artifacts contained less than 1.86 percent zinc and are considered copper (see Ehrhardt 2002:Appendix E). Of these, the compositions of two, the bipointed piece and the awl, contained such low trace element values as to be considered native copper. Both artifacts have very low levels of silver, arsenic, iron, and mercury. Their levels of arsenic (As), lead (Pb), tin (Sn), and antimony (Sb) fall below PIXE's typical detection limits for these elements (As < .013 percent, Pb < .040 percent, Sn < .034 percent, and Sb < .038 percent respectively [Fleming and Swann 2000:710]).

The levels of nickel (Ni) just slightly exceed these limits (< .017 percent). This confirmed hypotheses generated from the visual inspection that these two objects were made of native copper.

European-Introduced Copper

Copper content of the European-derived copper artifact samples ranges from 95.5 percent–98.3 percent Cu. Of the diagnostic elements that were measurable,[2] lead and antimony are found to be the major impurities. The group also features a low tin (< .50–.80 percent Sn), higher tin (.138–.232 percent Sn) and high tin (1.086 percent Sn) distribution. Hancock et al. (1995) have found similar tin-rich European-derived copper artifacts on native sites in Ontario. No similar enrichment grouping patterns were noted for arsenic, which ranged in content from below detection limits to .294 percent. The highest arsenic value (.294 percent As) belongs to a small, flattened arsenic-rich ring made of single-lobed tubing.

Notable among compositions of individual copper artifacts is that of the tiny ring (see Figure 6.8 b) manufactured of single-lobed "e" shaped tubing. Compared to the other artifacts in the smelted copper group, the tubing is relatively high in tin (.189 percent Sn) and has the highest concentration of many diagnostic trace elements, including arsenic (.294 percent As), lead (1.217 percent Pb), silver (.380 percent Ag), nickel (.606 percent Ni) and antimony (.934 percent Sb). Its arsenic value is the highest in the entire subsample. Metallographic analysis confirmed that this is a complex alloy in a precipitate phase. None of the other artifacts made of hollow tubing follow this compositional pattern.

Brass

The remaining 34 artifacts contain higher levels of zinc (14.66 percent–32.63 percent), indicating purposeful alloying to produce brass. Within this group, three artifacts, a rolled bead, a blank fragment, and a piece of rectangular scrap are classed as "red brasses" (in modern industrial terminology) as each contains about 15 percent zinc (14.66 percent–15.38 percent) (Brady et al. 1997:120). All of these red brasses also contain between 3.11 percent and 4.7 percent tin. Even small additions of tin to brass are known to increase the metal's hardness but reduce its ductility (Brady et al. 1997:119,126). It also increases resistance to corrosion. In addition, and perhaps importantly here, as George Brady et al. (1997:120–121) point out, "pleasing colors" are also produced when tin is added to brasses low in zinc (see also Day 1998:136). William

Fitzgerald and Peter Ramsden (1988:157) have commented that those brass artifacts in their sample with greater than 1 percent tin predate A.D. 1600.

One of the red brass artifacts in the Iliniwek sample, the rolled bead, also contains elevated levels of lead (1.451 percent). Lead added to brass in amounts between .5 percent and 3.0 percent is known to make metal more suitable for machining (Untracht 1968:18).

Thirty-one artifacts contain between 22.47 percent to 32.63 percent zinc, a more typical range that includes modern cartridge brass (nominally 70Cu–30Zn). These high-zinc brasses are strong and have excellent ductility. In modern metalworking, these brasses are known for their working qualities in drawing, spinning, and stamping processes (Brady et al. 1997:120).

One other compositional finding concerning the brass in the Iliniwek Village sample deserves attention. The observed range for lead (Pb) is noteworthy. As mentioned earlier, purposeful additions of small quantities of lead improve brass's machinability and its suitability for other types of working (Brady et al. 1997:120). From .5 percent to over 3.0 percent lead may be added. Among the smelted copper artifacts, only five (13 percent) have lead levels over .5 percent. However, 21 brass specimens (62 percent) have lead contents in excess of .5 percent; two beads possess 1.5 percent lead and one blank contains almost 2.0 percent lead. Lead appears in much higher levels in the brass from the subsample than in the copper.

While these findings may appear to indicate that lead or other metals (tin, for example) are being added purposefully (which might be expected), such a conclusion may be premature (Day 1998:133). During this period, inhomogeneity of smelted metal, especially copper alloys, is well known. The highly variable lead contents, notably in the brass artifacts, demonstrate that phenomenon (Pollard and Heron 1996:211). However, it must be kept in mind that much of the compositional inhomogeneity seen in brass artifacts of the period may well be due to the fact that metalworkers often remelted old objects and/or mixed scraps into their "melting pots" resulting in wares with quite dissimilar compositions depending on the particular batch of metal from which they came (Day 1998:134).

The Material Character of the Subsample

PIXE compositional results were surprisingly consistent in terms of the relative frequency of European-derived copper and brass in each typological category (Table 7.1). Overall, these materials are found quite evenly distributed in the manufacture of both finished and unfinished items. While there are more

Table 7.1 Proton-Induced X-Ray Emission Spectrometry (PIXE) Results of European-Derived Copper-Base Metal Artifacts by Artifact Type

Type	Copper	Brass
UNFINISHED ARTIFACTS		
Blanks (19)	●●●●●●●●	⊖○○○○○○○○○○
Tubing (9)	●●●●●●●●	○
Scrap (21)	●●●●●●●●●●●	⊖○○○○○○○○○
FINISHED ARTIFACTS		
Beads (rolled) (7)	●●●	⊖○○○
Bead (spiral strip) (1)		○
Clips (11)	●●●●●	○○○○○○
Tinkling Cones (4)	●●	○○
Ring (tubing) (1)	●	

Key:
● copper
⊖ red brass (14.66–15.38% zinc)
○ cartridge brass (22.47–32.63% zinc)

brass blanks in the sample than copper ones, there are also more finished items made from brass blanks than from copper blanks. Single- and double-lobed tubing stands out as the only category in which there is a clear preference for one material over another. In this instance, copper is most likely selected for this tubing, undoubtedly because of its superior ductility in such delicate shaping.

These findings are significant in terms of availability and native preference. Aside from selecting copper for the manufacture of tubing, PIXE compositional evidence suggests that there is no preference for one material over another for producing particular artifact forms. Beads, clips, and cones of both European copper and brass appear in relatively equal proportions.

These results are also interesting with regard to Fitzgerald's (1995:35) conclusions concerning the relative availability of European trade copper and brass after 1600. He notes an increase in brass material in inland native scrap assemblages after that date, and hypothesizes that a shift in supply from more elaborate copper kettles to less expensively produced brass containers caused the trend. This is not borne out at the Iliniwek Village. The relatively even occurrence of European-introduced copper and brass would indicate more balanced availability.

This is not to assume that all of the metal found at the Iliniwek Village comes from the same source or was obtained at the same time. In addition to the compositional variation noted in the artifacts, the thickness of the metal in the subsample ranges from .2–.8 mm. This provides some evidence that it is not all from the same stock. In the overall sample, a few artifacts are as thick as .7–.8 mm (see Figure 6.10 b), and the thickness of one piece of scrap even exceeds 1.0 mm (see Figure 6.10 c). While these occurrences are rare, it does argue for a somewhat varied raw material repertoire.

At the same time, however, it is interesting that two copper tinkling cones of the same "open midsection" style (see Figure 6.6 a, left and second left) and from the same feature (F136) are made of metal of the same thickness (.5 mm). Compositionally, their copper and iron contents compare very favorably but their lead and silver counts differ. Nickel, antimony, and tin were all below detection limits, so appear to be negligible impurities. By the same token, two copper rectangular blanks from the same feature that were thought to "refit" were also analyzed as part of the subsample. Their elemental values by PIXE are quite similar. Whether they are actually made from the same piece of metal is not known on the basis of PIXE results.

RESULTS OF THE MICROSTRUCTURAL (MEGALLOGRAPHIC) ANALYSIS

Sixty-four of the 75 artifacts in the subsample were examined metallographically. Sixty-two of the artifacts were either European-introduced copper (n = 31) or brass (n = 31). The two remaining samples were native copper. Examples of all major artifact forms in the subsample were studied.

Interpreting the technological processes involved in fabricating the artifacts is based on combined evidence from 1) PIXE compositional analysis, 2) visual and light optical microscopic examination, and 3) the patterns of work history revealed in the microstructures of unfinished and finished artifacts. Some general remarks about the metallographic microstructures precede their technological interpretation. In general, the microstructures revealed moderately to heavily corroded artifact surfaces. Corrosion was primarily intergranular in nature. Attack was often uneven, with varying degrees of pitting noted along a single specimen. Corrosion cracking was evident on some pieces, and in several cases, pitting penetrated the entire thickness of the piece. Corrosion preferentially attacked cold worked areas of the specimens. While alloying purportedly improves corrosion resistance, there appeared to be no appreciable difference in degree of corrosion from copper to brass (Newman and Sieradzki 1994:1708; see also Cushing 1967 for discussion).

Micrographs of European smelted copper-base metals showed varying numbers and types of inclusions (see Figures 7.1 and 7.4 for examples). The inclusions appear to be duplex in nature and are commonly made up of sulfides, silicates, and oxides. Lead inclusions also occur. Inclusions are found both within the grains and at the grain boundaries. They vary in size and shape within and across specimens. They are globular and/or elongated. Elongated inclusions are either dispersed or strung out to various degrees indicating the directionality of working. The two native copper specimens were virtually inclusion-free.

As shown in chapter 6, some degree of bending is apparent on all artifacts in the sample. These operations include (but are not limited to) shearing, folding over flush, twisting, or simply shaping into a finished form by rolling or coiling the metal freehand or around some sort of form or core. The degree to which these actions are revealed in the microstructures depends on the extent to which the bending operation deforms the metal.[3] As O. D. Lascoe (1988:181) has said, "Bending of sheet metal has the distinct characteristic of stressing the metal at localized areas only. This localized stress or pressure occurs only at the bend radius. The remaining or flat metal is not stressed during bending." This pattern is apparent in the microstructures of the artifacts in this study.

As will be seen below, the mild bending operations required to form copper or brass sheet into final shape in the way it is frequently seen in this sample are often not sufficient to cause a perceptible change in the microstructure. Overall, much of the bending actually consisted of rolling an annealed sheet into desired shape. These actions may deform the metal but this degree of bending deformation is normally not visible. However, shearing, as one form of bending, does cause localized deformation that is visible microstructurally. Effects of this operation are readily and frequently visible on the margins of these samples.

TECHNOLOGICAL INTERPRETATIONS OF THE MICROSTRUCTURES

European-Derived Copper-Base Metals—Unfinished Artifacts

Of the 42 artifacts in the "unfinished" or "partially processed" group, 23 (55 percent) are copper and 19 (45 percent) are brass. In general, their microstructures reveal that these artifacts were left both in an annealed condition

Table 7.2 Summary of Metallographic Results

Type	Annealed	Annealed/previous cold working	Work hardened
EUROPEAN COPPER/BRASS - UNFINISHED ARTIFACTS			
Tubing (6)	●●	●●●	○
Blanks (18)	●● ○○○○	●●●	●●⊖○○○○○○
Scrap (18)	●●●● ○	●●●●●● ○○	●⊖○○○
EUROPEAN COPPER/BRASS - FINISHED ARTIFACTS			
Bead (spiral) (1)			○
Ring (tubing) (1)		●	
Tinkling Cones (4)	○	●●	○
Beads (rolled) (6)		●●⊖○○○	
Clips (8)	● ○○○	●● ○	○
NATIVE COPPER ARTIFACTS			
Awl (1)			●
Bipointed Object (1)		●	

Key:
● Copper
⊖ Brass containing 14.66–15.38% zinc
○ Brass containing 22.47–32.63% zinc

and in a wide range of deformation stages, indicating that they had been submitted to varying amounts of manipulation before they were discarded or abandoned (Table 7.2). Frequently, gradations of deformation are seen on a single microsection, indicating random, irregular, or differential working or variation in annealing temperature along a particular piece. Because the pieces are so small, it is likely that working varied along the piece.

Overall, 27, or 64 percent, of the unfinished artifacts were left in an annealed condition, that is, they exhibited a twinned, equiaxed grain structure (Figure 7.1). Twenty are copper; the remaining seven are brass. All categories of unfinished artifacts are represented. Exactly half of the blanks, all but one of the pieces of tubing, and 13 of the 18 pieces of scrap were left annealed. Of these, 11 exhibit a uniform medium or coarse grained, twinned, equiaxed grain structure, indicating no previous cold working. Two possess variable, nonuniform grain sizes that look to be the result of a high temperature anneal rather than previous cold working.

Over half (n = 14) of the partially processed objects left in an annealed

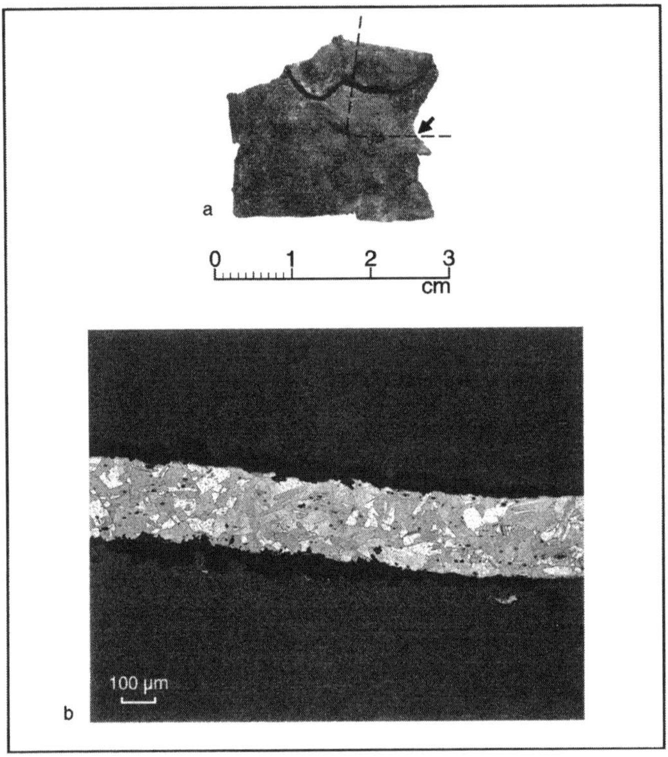

Figure 7.1 Processing unfinished artifacts: annealing: (a) irregularly shaped piece of smelted copper scrap, cut, bent, folded, one fold possibly hammered down; (b) the section of the artifact examined was left in an annealed condition, without indication of previous cold working. Typical twinned, equiaxed grain structure. Bichromate and ferric chloride etch. 100x.

condition show clear, often multiple signs of previous cold working (Figure 7.2). These indicators include small or variable grain sizes, transgranular cracking, laps/seams, elongation of inclusions, and/or preferential corrosion attack in previously worked locations. For the most part, the elongated, strung out inclusions noted on some pieces result from the initial manufacture of the sheet. In one large trapezoidal blank (pictured in Figure 6.9 b, center), however, the size and shape of the inclusions change over the length of the piece indicating differential and probably quite heavy cold working *after* initial manufacture. Small and/or variable grain size, which most often occurs on the edges of the specimens, is due either to localized, differential cold working, or to deformation related to shearing (see the inside edges of Figure 7.2). One

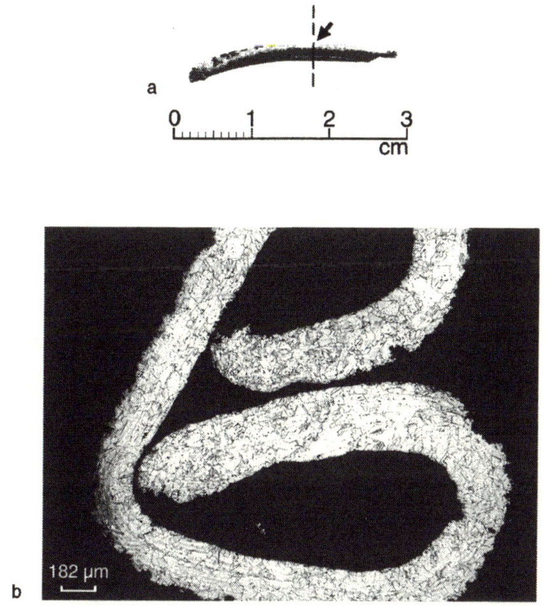

Figure 7.2 Processing unfinished artifacts: annealing with prior deformation: (a) segment of double-lobed smelted copper tubing; (b) microstructure shows that this artifact was left in an annealed condition after having been deformed. Not shear burrs and prior differential deformation at ends due to shearing. Bichromate and ferric chloride etch. 55x.

piece of scrap left in this condition had been folded over and hammered so heavily prior to annealing that it had been reduced to the thickness of one piece of metal. Most importantly, the occurrence of this feature indicates that these artifacts were annealed *after* the sheets were cut and/or differentially cold worked. This means that these artifacts, which all showed evidence of working of some type—folding, hammering, shearing—were subjected to a final anneal before abandonment. Why this should necessarily have happened is unknown.

The microstructure of one artifact, a piece of folded copper scrap, is enigmatic (Figure 7.3). It is a rather extreme, but revealing example of the multiple manipulations some of these artifacts actually underwent before abandonment. Although the scrap was left in annealed state, it clearly had been heavily cold worked and hammered extensively. Long, unbroken, elongated inclusions

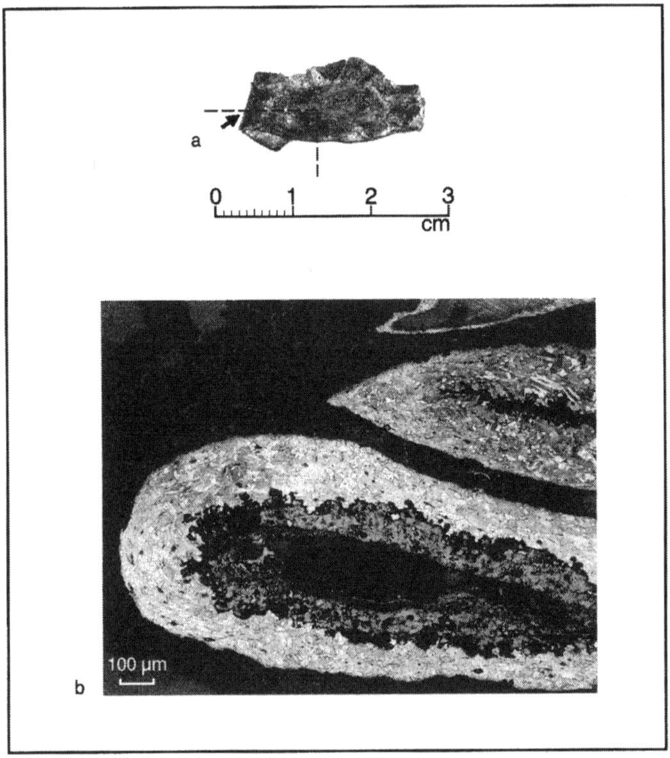

Figure 7.3 Processing unfinished artifacts: hot and cold working, annealing: (a) irregularly shaped piece of European copper scrap, bent and folded repeatedly; (b) specimen was hot worked, then differentially cold worked, then annealed. Scrap was left in an annealed condition. Bichromate and ferric chloride etch. 100x.

in the microstructure indicate that it had been hot worked beforehand. Initial hot working, subsequent cold working, probably including folding and hammering, then annealing slightly above the recrystallization temperature produced the small grain size on the current structure. Owing to the variation in the degree of subsequent cold working across the piece, some sections along it, particularly on the folds, have extremely fine grain structure and others have a more coarse structure.

The remaining 15 unfinished artifacts (36 percent) had been cold worked to some extent and were abandoned in that state. Twelve are brass and three are copper. Again, all typological categories of unfinished artifacts are represented. Half of the blanks, but only 26 percent of the scrap, and only one piece of heavily hammered, flattened brass tubing were left cold worked (Figure 7.4 c). In contrast, all of the copper tubing was left annealed. Few of the specimens were uniformly cold worked ($n = 3$); these include the single piece of flattened brass tubing. Most exhibited uneven gradients of deformation across the section ($n = 7$) or a single zone of cold working on an otherwise annealed structure ($n = 5$) (Figure 7.6 is an example of this). Examples from both the "scrap" and the "blanks" category carry these characteristics; most of the pieces are made of brass.

Plastic deformation in all of these samples is likely attributable to bending, shearing, or hammering with a hammerstone or other tool. The degree of deformation varies, ranging from light to intermediate, to advanced/heavy (Vernon 1990). Localized heavy cold working is clearly associated with folding and hammering down folds, flattening tubing, or thinning to a point. Some shear burrs are also found to have been hammered down quite heavily.

While almost two-thirds (64 percent) of the unfinished artifacts are left in an annealed condition, there are distinct differences in the pre-abandonment treatment of copper and brass. Brass artifacts are most frequently left work-hardened (63 percent), whereas 87 percent of the copper is abandoned in an annealed or deformed then annealed condition. Brass is left cold worked over four times as frequently as copper. Brass blanks and scrap are preferably work-hardened, whereas copper blanks and scrap were left annealed.

European-Derived Copper-Base Metals—Finished Artifacts

The microstructures of the finished objects ($n = 20$) stand in interesting contrast to unfinished artifacts and provide valuable insight into final shaping and finishing processes (Table 7.2). Of the 20 finished artifacts, eight are copper and 12 are brass. As in the case of the partially processed artifacts, the majority of the finished artifacts were left annealed. Indeed, 17, or 85 percent, were left in this condition. However, in contrast to the unfinished artifacts, which were distributed fairly evenly between the annealed (48 percent) and annealed with previous cold working (52 percent) categories, an even greater proportion (71 percent) of the finished artifacts are seen to have undergone some degree of cold working prior to annealing (Figure 7.5). These results clearly indicate that prepared blanks or preforms were worked then given a final annealing treatment to restore the metal to a softened, ductile, and malleable state. This may well have been done in order to facilitate final shaping, particularly to prevent the kinks or sharp bends sometimes created by trying to roll or coil work-hardened or even embrittled metal (Vernon 1990). The fact that so

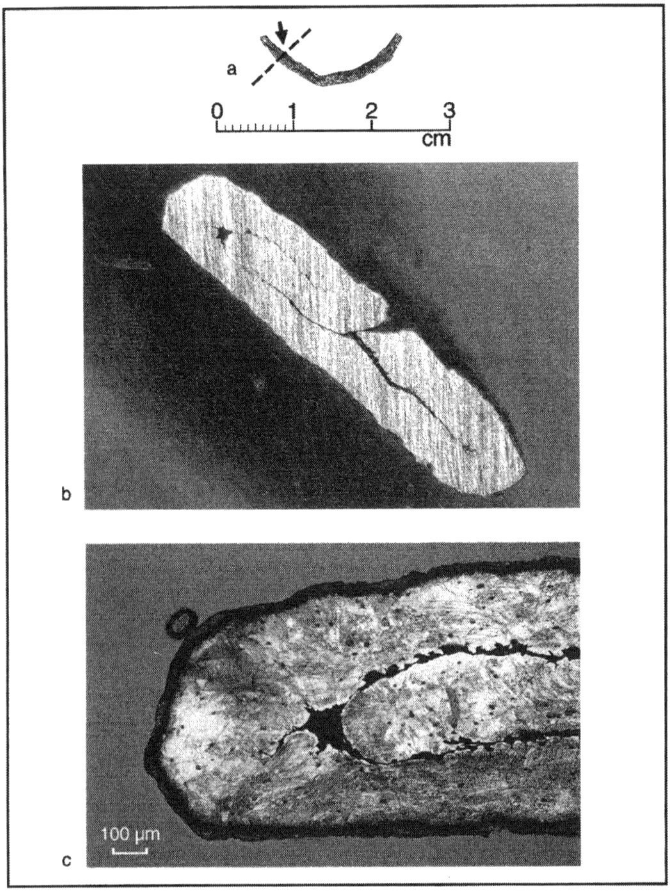

Figure 7.4 Processing unfinished artifacts: heavy cold working: (a) flattened segment of double-lobed *brass* tubing; thin (.8 mm) and looks to have been hammered heavily; (b) unpolished, unetched section. 36x; (c) specimen was left in an advanced stage of deformation. The microstructure shows the heavily distorted grains along with flow lines indicative of heavy cold working. Bichromate and ferric chloride etch. 100x.

many of these artifacts were left in an annealed instead of a work-hardened condition is consistent with the notion that as ornaments, they did not have to function as working, utilitarian tools (Vernon 1990).

Interestingly, more copper artifacts were cold worked, then annealed (n = 7) than were brass artifacts (n = 5). Most of this kind of processing in both materials appears to have been carried out on rolled beads, but primarily on brass beads. The bulk of the evidence for cold working comes from the mar-

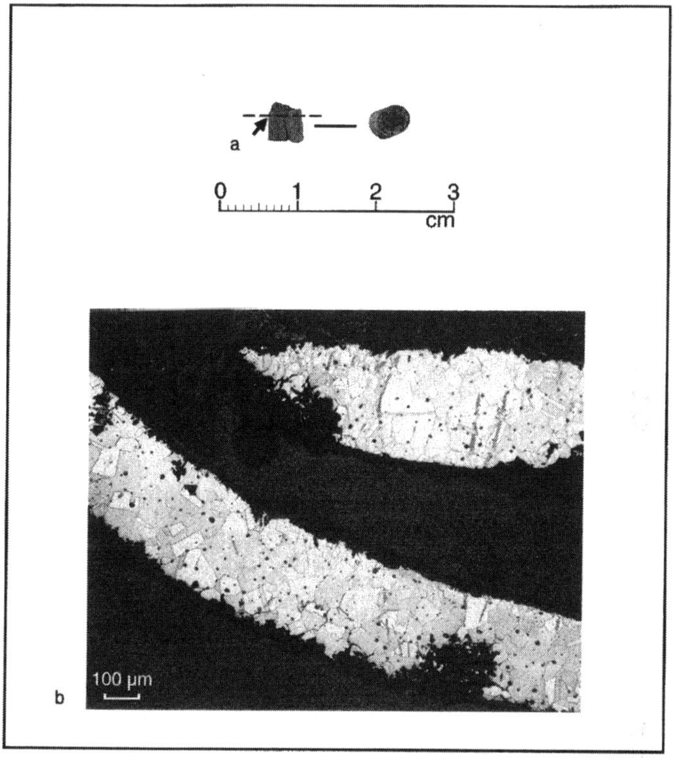

Figure 7.5 Forming finished artifacts: annealing: (a) brass bead with overlapping closure; (b) this specimen was left in an annealed condition, and shows slightly smaller grain size and differential corrosion at inside closure, indicating deformation in that area prior to the final anneal. Bichromate and ferric chloride etch. 100x.

gins of these pieces, which had been distorted in the shearing process (Figure 7.5). This is indicated by the smaller grain sizes noted on the margins of these specimens. The differential corrosion present on the closure of one of these brass beads may indicate some cold working (perhaps hammering) at that location but the corrosion now obliterates it. One of the copper beads, however, had undergone extensive cold working, as its microstructure reveals a seam near one short end. Grain size at the end of the blank is 6 1/2 whereas in the midsection it is 9 1/2. In any case, these artifacts were annealed *after* these processes took place. Whether this happened before or after bending into shape is not known. Importantly, no detectable deformation took place after the anneal.

Only three of the finished artifacts, a clip, a tinkling cone, and the spiral strip bead, were left work-hardened. Importantly, all are brass. In the case of the clip (illustrated in Figure 6.5 a), annealing may have been followed by a "skinning" (light cold working) treatment administered to smooth out rough, sheared edges, as slight strain markings are visible on the lateral margins of the microstructure (see Ehrhardt, Nash, and Swann 2000:Figure 4 c).

The light deformation on the outside of the neck of a tinkling cone (Figure 7.6 b) likely results from some postforming spot hammering applied to close its neck seam flush. No finished artifact exhibited any more intense cold working than this type of light treatment.

The manufacture of the spiral strip bead is the only exception to the shaping sequence revealed above in which an anneal was performed to facilitate the actual forming process. The moderate shearing deformation present along the edges of the strip blank indicates that this object was coiled in a lightly cold worked state (Ehrhardt, Nash, and Swann 2000:Figure 4 d). No kinks or yield points are noted on the finished artifact, however.

Native Copper Artifacts

Microsections of the two native copper artifacts in the industry are illustrated here (Figures 7.7 and 7.8). The microstructure of the small awl (Figures 7.7 b, c) reveals that it was formed by extremely heavy and repeated cold working, folding, and annealing (after Franklin et al. 1981) of a nodule of native copper. It shows a lap (see Chapter 6 endnote 6) and an internal burst (Figure 7.7 c) indicative of this type of treatment. Left in a work-hardened condition, deformation is extremely heavy from the tip (or the working end) through the upper body (Figure 7.7 b), as would be expected if the object were meant to function in a utilitarian context.

The remaining native copper artifact is a very small, ovoid, bipointed piece that fits no recognizable formal category (Figure 7.8). Likewise, its function is unknown. Unlike the awl, it was left in an annealed condition. However, the grain size is small and variable, with the smallest grains found toward the outside margins of the piece. The pattern of grain size variability and the presence of a lap indicate that heavy cold working preceded the anneal.

INSTRUMENTAL NEUTRON ACTIVATION ANALYSIS (INAA)

Chemical Composition of the Artifacts by INAA

As outlined in detail in chapter 4, INAA has figured prominently in archaeometallurgical studies of protohistoric/early contact period metals composi-

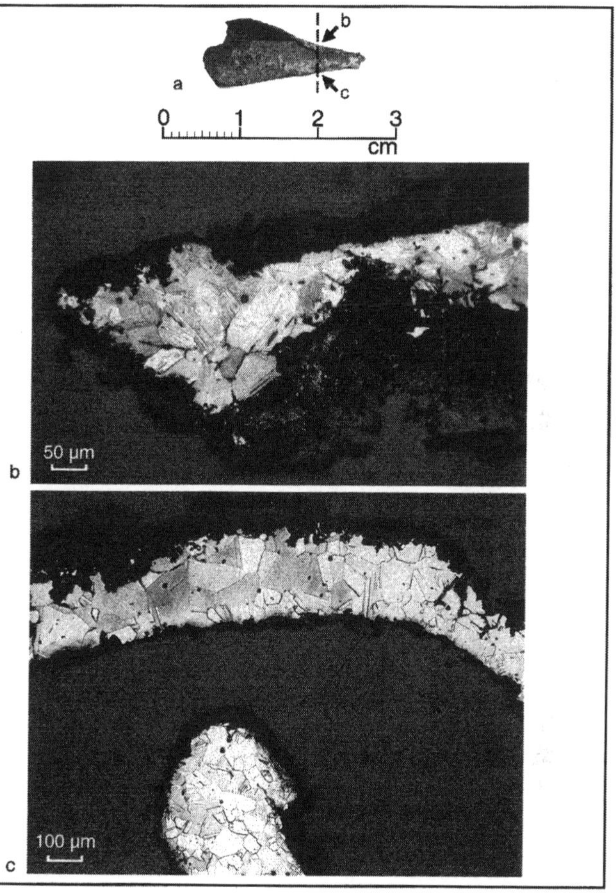

Figure 7.6 Forming finished artifacts: differential light cold working: (a) brass tinkling cone with an oblique depression across the neck closure; (b) light to moderate deformation occurs on the *outside* of the neck closure, whereas, the *inside* (c) shows no similar pattern of grain deformation. Therefore, the working on the outside took place *after* the cone was rolled. The closure was probably hammered or squeezed down after bending. Bichromate and ferric chloride etch. (b) = 200x; (c) = 100x.

tion, consumption, distribution, and use. In addition to its utility in distinguishing native copper artifacts from those fashioned of European-introduced copper and brass, INAA results have been used to sort similar individual chemistries of European-introduced copper-base metals into like compositional groups. These groups are interpreted to reflect particular "lots" or "batches" of kettles. While these studies are promising for exploring intra- and intersite distribution and consumption of trade metals, tracing groups' move-

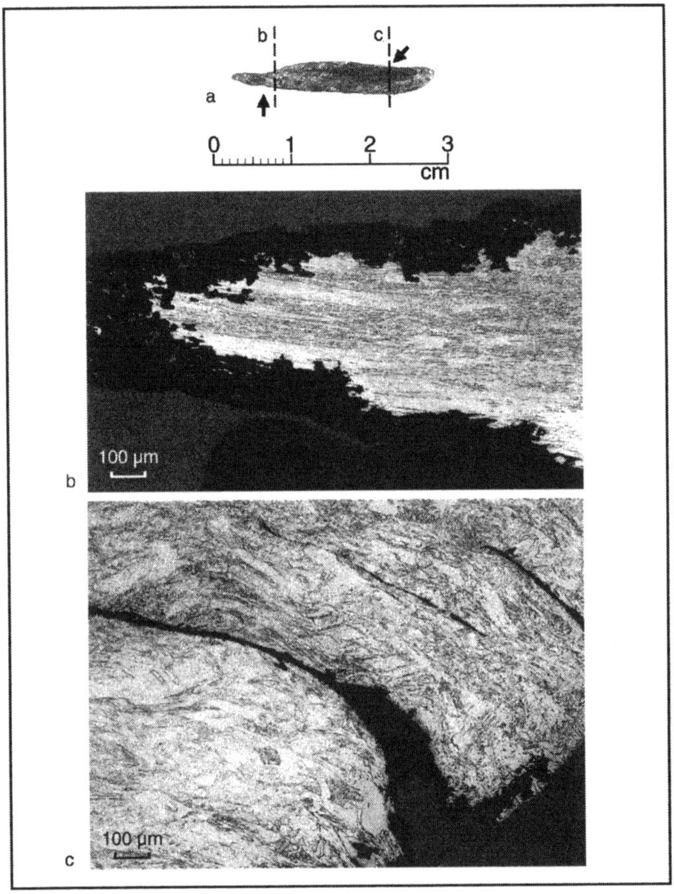

Figure 7.7 Forming native copper: (a) native copper awl; (b) longitudinal tip section reveals an extremely heavily cold worked structure. Flow lines and strain markings dominate. Bichromate and ferric chloride etch. 100x; (c) the body of the awl is also heavily cold worked. The metal has been folded during the forming process. Laps and internal bursts are indicative of extreme cold working. Note the absence of inclusions in native copper. Bichromate and ferric chloride etch. 100x.

ments, or even for identifying patterns of interregional trade, the data generated in this study is not analyzed here for that purpose except in a limited and generalized way. Results remain suggestive rather than conclusive.

Seventy-two samples from the 75-artifact subsample were submitted to chemical analysis by INAA (Ehrhardt 2002:Appendix F). Three segments of hollow tubing were omitted from INAA because they were too heavily cor-

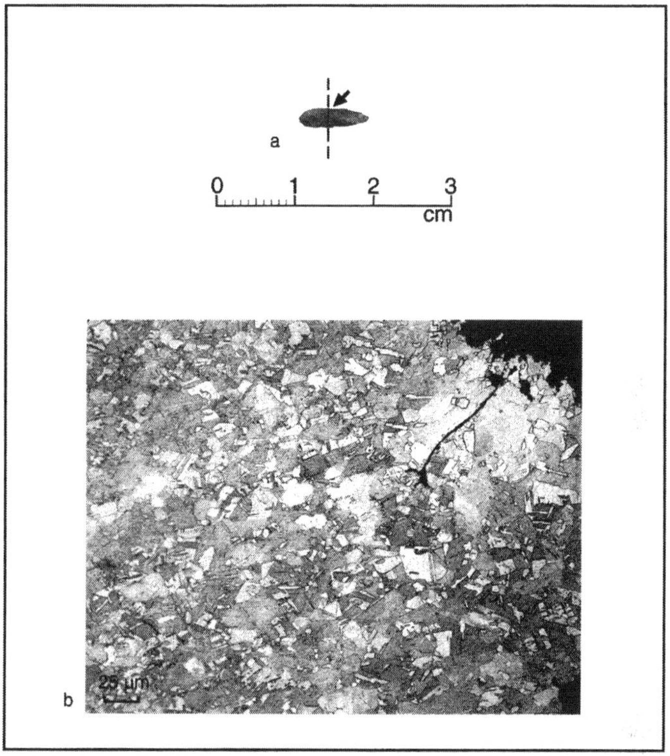

Figure 7.8 Forming native copper: heavy cold hammering, folding, annealing: (a) small bipointed native copper artifact of unknown function; (b) this artifact was left in an annealed state but prior cold working is apparent. The metal has been folded and carries a seam generated in the cold working stage. Bichromate and ferric chloride etch. 400x.

roded; too little clean metal remained to obtain a reliable reading. Two samples were taken from five of the specimens, bringing the total to 77 samples.

Twelve elements, copper, zinc, tin, arsenic, iron, silver, nickel, antimony, gold, cobalt, selenium, and tellurium, were measured in this study. Nine of these had also been measured by PIXE. Since INAA detection limits were generally lower for zinc, absolute concentrations of this element were reported in all but eight samples in which the zinc content was extremely low. However, INAA did not detect tin values below 132 ppm. The iron content of all of the copper (both European and native) samples was also below INAA's detection limits.

INAA results were comparable with those derived from PIXE in terms of major source categories; 38 artifacts were determined to be copper (containing less than 1 percent Zn) and 34 were brass. The two native copper artifacts were easily identified. Most elements except cobalt, silver, arsenic, and tellurium appeared in such low concentrations in the native copper artifacts that they did not reach detection limits. Only silver was measurable in both artifacts.

Among the European copper artifacts, zinc concentrations range from below detection limits to 537.03 ppm (or .05 percent). This does not constitute purposeful alloying. Within the European copper category, tin levels are again seen to group into low (below detection limits), middle (132.25–839.05 ppm), and tin-rich (1264.95–6443.61 ppm) ranges. Arsenic levels vary widely (5.050–2174.92 ppm) as do those of silver (153.53–2094.18 ppm). Antimony registers the widest range of trace element concentrations within the copper category (18.22–4814.03 ppm). However, in this sample, antimony concentrations are generally much higher in the copper category than they are in that of the brass; 26 of 38 artifacts in the copper group have higher trace element values of antimony than any of the brass artifacts.

Based on the similarities of their chemical compositions as measured by INAA, there are three "sets" of two artifacts within the European-derived copper group that are thought to come from the same piece of metal or at least from the same batch of metal. Criteria for making this determination come from Hancock and colleagues who have found that trace element content of the sheet metal within individual kettles varies between ± 10 to ± 20 percent (Hancock, Fox, Conway, and Pavlish 1993:311; Hancock, Pavlish, Fox, and Latta 1995:336–347). The compositional fingerprints of the two tinkling cones (mentioned earlier) fall within that range (see Figure 6.6 a extreme and second left), the two blanks that "refit" on visual inspection (also mentioned above), and another group of two beads. While the two beads and the two blanks come from the same feature, their chemical compositions are not similar. INAA, then, confirms the hypothesis that the first two groups suggested to be affiliated by PIXE were valid and reveals another. It is interesting to note that all three groups are low in antimony.

The trace element concentrations seen for the tubing from which the small ring was made as measured by PIXE do not compare favorably with the INAA assay. While its tin and nickel concentrations by INAA remain very high, silver is now found near midrange. Interestingly, its arsenic content remains among the highest in the entire subsample, and is compatible with those of the "red brasses."

Compositional results were less consistent in the brass category, however.[4] Zinc contents ranged from 157170.4 ppm (15.7 percent) to 340631.56 ppm

(34 percent). A "red" brass range was identified. It contains the three artifacts placed in this category based on PIXE results plus another that registered as having 15.7 percent zinc (157150.4 ppm) by INAA but 26 percent zinc by PIXE. In addition, all four were found by INAA to have correspondingly high tin (over 3 percent) concentrations, a feature that also accompanied the three red brasses identified by PIXE. Interestingly, however, all four of these artifacts remain consistent compositionally as they have the highest arsenic, nickel, antimony, and iron concentrations in the brass category. At the same time, they have relatively high cobalt and low gold (9.89–11.84 ppm) values. Their silver concentrations are inconsistent but fall in the mid-to-high range. Indeed, two of these artifacts, a rectangular blank and a piece of rectangular scrap, are from the same feature and have quite similar chemical fingerprints. While slightly out of the allowable variability range (± 10 percent – ± 20 percent) in some elements, one might hypothesize (based on INAA results alone) that they came from the same piece or batch of metal. However, when lead is considered, PIXE revealed that the former contains 1.924 percent lead, whereas the latter has only .667 percent.

The remainder of the brass artifacts falls more squarely into the cartridge brass category. Tin levels among these high-zinc brasses did not reach the levels recorded for the low-zinc brasses, barely reaching one-third the concentrations found in that group. Arsenic levels remain relatively low (below 440.00 ppm) in the high-zinc brass group as well. Outside of the high levels found in the low brass group, only two, a blank and an ovoid bead have arsenic contents falling within a midrange (723.58 ppm–904.00 ppm respectively).

Compositional Grouping

Although the data set is small for this purpose, preliminary attempts were made to "group" these individual brass and copper chemistries into distinct, statistically valid "lots" or "batches" of metal with like compositions. Protocols established at MURR for recognizing patterns and validating them statistically were followed here (Neff 1994). Simultaneous R-Q mode principal components analyses were performed on the log-base 10 and Aitchison-transformed compositional data to detect these patterns. The analyses were done with and without copper values. Iron and tellurium were eliminated from the copper group analysis, and tellurium was not used in the brass analysis. Some preliminary results are presented here.

The investigation met with varied success depending on the copper-base metal category under consideration. The copper data set separated fairly discernibly into four groups with several samples remaining ungrouped (Figure 7.9). Group 1 is a high silver/gold/cobalt group with low arsenic and nickel. Group 2 is a high antimony/arsenic group with low silver, gold, and cobalt. Group 3 is midrange in many elements. Group 4 is generally higher in zinc, very high in tin, and low in cobalt. Thus, antimony, tin, zinc, and cobalt all played significant roles in differentiating the compositional groupings. Tin covaried with antimony. Selenium was influential as well. Bivariate plots were used to further inspect the data and to test the compositional groupings tentatively identified in the principal components analysis.

Brass was significantly more difficult to group (see Hancock, Farquhar, Pavlish, and Finlayson 1995 for similar problems grouping brass). Once the low or red brasses were removed from the data set, outside of tin and cobalt, no single element accounted for significant variation in the remaining brass sample (Figure 7.10). While tin ostensibly influenced the groupings most significantly, gold, silver, copper, zinc, arsenic, and cobalt all contributed to variation. While groupings based on high cobalt/arsenic values and on low antimony or high cobalt values were suggested, no sustainable groupings could be discerned or confirmed by Mahalanobis distance-derived probabilities of group membership. The metallurgical reasons discussed above are likely responsible for this. In seventeenth-century foundries, metal production, particularly alloying operations, were hardly standardized as they are today. Waste pieces of metal could have been thrown back into the pot, causing adulteration of individual melts by inclusion of various waste or recycled products, therein muddying the possibility of detecting distinct source related (or supply related) groupings (Day 1998).

As stated previously, INAA was highly successful, however, at detecting sample-to-sample affinities. In four cases, elemental concentrations of two artifacts deposited in the same pit fill (but not necessarily of the same type) were so close compositionally (within ± 10–20 percent after Hancock, Pavlish, Fox, and Latta 1995:346) as to suggest strongly that they were fashioned from the same piece or batch of metal. In one of the cases, compositional data verified a suspected refit of two rectangular blanks from within a cache of over 35 such objects. How many other artifacts from that provenience may have matched compositionally can only be suggested.

ILLINOIS COPPER-BASE METAL ARTIFACT PRODUCTION—METALWORKING "STYLE" EMERGES

This section integrates what has been learned from the various types of materials analyses conducted here about the manufacturing sequences and pro-

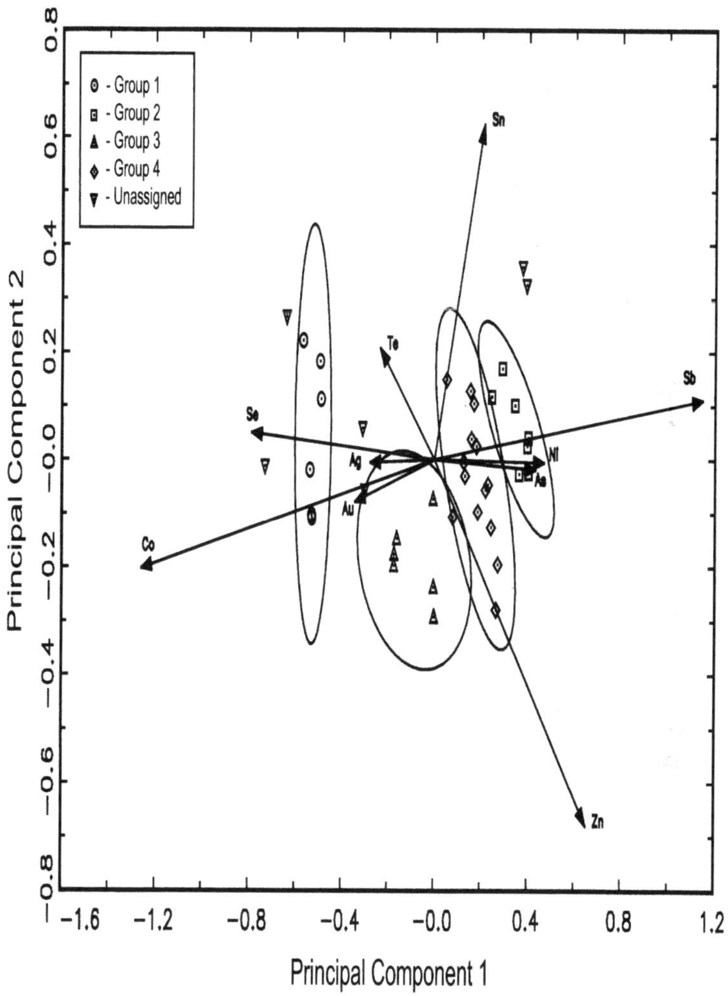

Figure 7.9 Principal Components Analysis of the log-base 10 and Aitchison-transformed smelted copper INAA data set (native copper has been removed). Preliminary groupings are shown.

cesses involved in copper-base metal artifact production among the protohistoric Illinois. Major formal categories as delineated in the overall sample are reviewed.

Rolled Beads

Small ovoid, round beads, and tubular beads are the most frequently found metal artifacts in the industry. Large tubular beads are absent. The small rolled beads vary in size according to shape. Beads are made of either European-

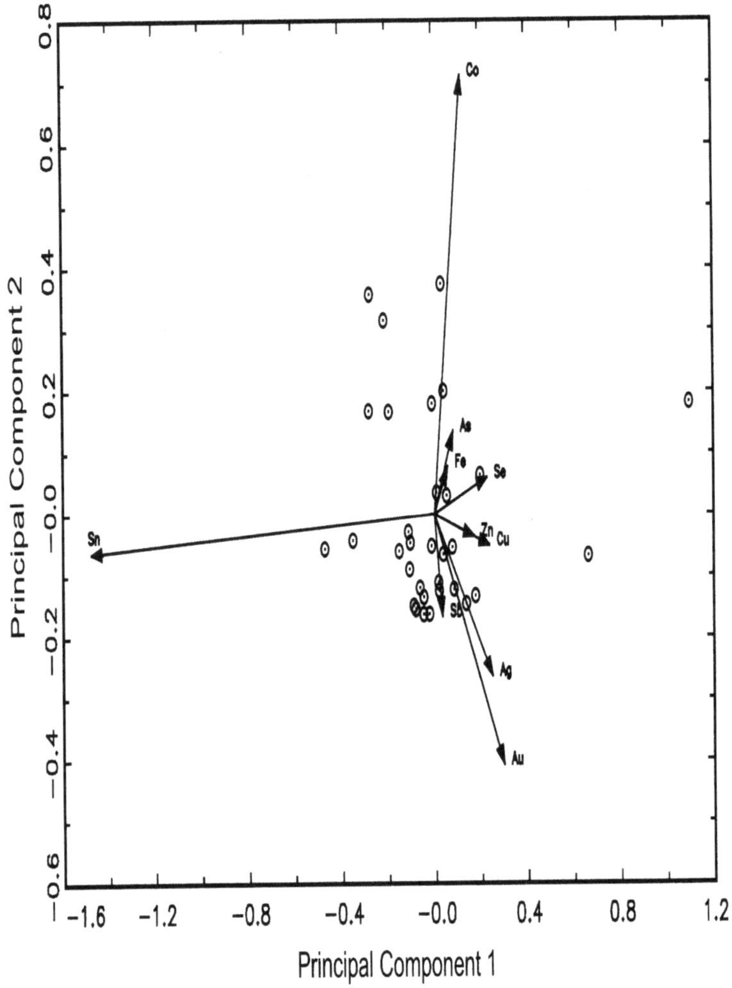

Figure 7.10 Principal Components Analysis of the log-base 10 and Aitchison-transformed high-zinc brass data set (red brasses have been removed).

introduced copper or brass. There are no rolled beads of native copper in the subsample. Of the seven beads tested compositionally ($n = 7$), four were brass, three were European copper. Whether this actually represents a preference for brass over copper in the manufacture of beads is not known. One is made of the low-zinc red brass, which may have had a slightly more yellow color than copper. This may have made it particularly attractive.

Beads are manufactured in a simple manner. Blanks of a prescribed size were likely sheared into *roughly* rectangular shape from kettle sheet metal. Sheet was sheared most commonly using shears or through use of some type of sharp-edged tool. From the deep and jagged burrs noted on some of the blanks, it is apparent that these tools were employed with little regard to the cutting ability of the edge. In many cases, the blank cutting sequence can be seen by the pattern of overlap on the shear burrs. To take their final shape, beads were rolled over a mandrel (possibly for stringing) or fastened around leather thongs. Most often, their closures overlap. Several are found abandoned in mid-rolling; there are numerous examples of incompletely rolled small beads and a single example of a tubular bead that had been abandoned well into the rolling process. Metallographic evidence indicates that both copper and brass beads were annealed *before* they were rolled, probably to remove any work hardening effects of the shearing and/or other cold working and to facilitate the rolling process. Several had been worked to some extent at the short ends before final annealing. The red brass bead had been particularly heavily worked in that area. No beads were cold worked after annealing. The metal was in a recrystallized condition when the beads were used.

Regrettably, even with the preservative action of copper salts, at Iliniwek the organic material to which beads were fastened or on which they may have been strung rarely survives. A number of unattached beads carry oblique impressions where some sort of tool had been used to press the seams of the beads closed, perhaps over the materials to which they had once been fastened.

Fortunately, however, there is some direct evidence for their ornamental use. Single and multiple beads of relatively large size were found encasing some sort of unidentified organic material. In several instances, this material also contained a very small rectangular ferrous rod or wire. Whether these functioned as a part of the fringe of a garment or accessory is not known.

Spiral Strip Beads

These spirals, which resemble slender wood shavings, are found most frequently about the crania in native burials of this period. They are often still found attached to the material they encased. The three spiral strip beads in the Iliniwek sample appear to be smaller versions or replicas of longer examples found elsewhere. Two of the three are similar; they are made of somewhat different thicknesses and textures of sheet metal than the stock most commonly seen at Iliniwek. The least well-made example is flattened and looks to have been fashioned of the more prevalent material.

Compositional examination of one of these spiral beads (see Figure 6.4 a) revealed that it is made of brass. However, metallographic investigation did not shed much light on the technological derivation of the long, shallow striations that run longitudinally along the lateral margins of the blank. This feature was unique on these artifacts. While there was no reason to suspect that prepared strips of this width and length were being brought to the New World by Europeans or prepared on this side of the Atlantic by a blacksmith, it cannot be ruled out. In contrast, if they had been left by a native metalworker, they would likely have been the residual effect of a grooving or scoring process that would eventually have caused the metal to shear. The length, directionality, and evenness of the striations argue against them being polishing or grinding marks.

Metallography did reveal, however, that the spiral bead had been sheared and was subsequently shaped in a cold worked state. It was left that way. The sheared edges left characteristic, localized moderate to heavy strain markings on the lateral margins of an otherwise typically annealed sheet. Whether the other spiral artifacts were made in this manner is unknown.

Clips

Clips differ formally from beads and were likely used to achieve a somewhat different decorative effect. Although there is no direct evidence from Iliniwek as to how they were used, there is every indication that they were probably fastened onto a backing of some sort and were meant to lie flush with the surface of the material upon closure. The blanks from which these objects are made do not differ in size from the blanks used to make ovoid beads. Rectangular blanks for clip use look to have been fashioned in the same manner. Again, few are perfectly rectangular. However, in two cases trapezoidal or irregularly shaped blanks are shaped into rectangular blanks by folding them in half into a rectangular shape and hammering their edges down flush. These double-layered blanks are then rolled into clips.

Clips are prepared for use simply by bending up their short ends, creating "legs" which are then inserted into slits cut in a backing material and bent shut in staplelike fashion. It appears as though they are frequently squeezed shut, most likely between the fingers. Ample evidence attests to their reuse.

Clips are made of either imported copper or brass. In compositional analysis of 11 clips, brass ($n = 6$) clips again outnumber copper ($n = 5$). Most of the seven clips examined metallographically show that annealing was conducted *after* the blanks were cut into shape. Evidence of prior cold working, notably but probably not exclusively due to shearing, was documented on the margins of many of these objects. Any signs of bending deformation at the fold of the double-layered clip is obliterated by preferential corrosive attack. One brass clip (see Figure 6.5 a), however, was lightly worked along its margins after it had been annealed. Of all of the clips examined metallographically, it is the only one left in a cold worked state; the remainder of the clips were used in a softened, recrystallized condition.

Tinkling Cones

At least six different "styles" of tinkling cones were identified in the Iliniwek sample. The variety of irregular and potentially expedient forms they take well demonstrates one idiosyncratic way the Illinois shaped both imported copper and brass material. Of the four tinkling cones analyzed archaeometallurgically, two are copper and two are brass. The two copper cones (see Figure 6.6 a, left and second left) are of the same open-midsection style and are thought to have been made out of the same "batch" of material. Evidence from their microstructures further corroborates this suggestion. The metal from both cones has the same density and shape of inclusions. Both were fashioned in a similar manner; they were cold worked differentially along the edges and their necks show preferential corrosion at the point at which the closures meet. Closures were probably hammered down before the pieces were given a final anneal. Whether they were manufactured expediently and at the same time is not known with certainty, but they were probably made of introduced copper from the same source and were clearly discarded in the same location (House 1, Feature 136).

The two brass tinkling cones analyzed are similar in form but they differ in size. One (see Figure 6.6 far right; Figure 7.6) shows localized cold working only on the *outside* edge of the neck closure. This differential cold working was performed *after* the piece was annealed and *after* it was formed. It is the only cone left in a work-hardened condition.

Tubing

The tiny, misshapen ring is the only finished artifact analyzed in the subsample fashioned of one of many types of fine hollow tubing found on native sites in the Northeast. The piece of "e" shaped tubing from which it is made stands out in composition from other tubing specimens and from other metals in the subsample. Tubing is found at Iliniwek in sheared segments or in finished form as bracelets, coils, and rings. In all, there are over 1.75 linear meters of this tubing in the sample. Almost half of it comes from finished artifacts. As discussed previously in this work, its source, its distribution, and particularly its manufacture have been a true technological enigma for researchers. Both the "o," "e," and "B" forms appear so finely and evenly turned that it is difficult for researchers to come to a definitive conclusion as to its source. It appears in so many sizes, or "gauges" that one could not say its production was "standardized" or that its distribution reflects the dissemination of one "batch" of this material from Europe. Extensive inquiries to historical metallurgists (by this author) for any leads on this form as a European production were not fruitful. Ethnohistorians and archaeologists have not seen it listed or described in any documents relating to materials brought in for trade. Careful microscopic examination did not reveal longitudinal striations associated with drawing. It was hoped that archaeometallurgical investigation would at last be able to settle the question.

The fact that eight of nine pieces (six "B" shaped and three "e" shaped) of tubing submitted to compositional analyses turned out to be smelted copper is not at all surprising. Copper's ductility and malleability are well suited to the manufacture of these shapes. However, the trace element values from these specimens do not form a pattern in any way; there are no two compositionally similar fingerprints. This provides further evidence that they do not represent a single "batch" of imported material.

Metallographic investigation revealed that many of the specimens were heavily corroded. Three samples were subsequently omitted from INAA because they were too heavily oxidized to yield meaningful compositional results. All were found to have been rolled in an annealed state, except for the single brass example that had been heavily hammered and flattened (postforming), as indicated by the moderate to heavy deformation witnessed throughout (Figure 7.4). Particularly in the cases of the "B" shaped tubing, the effects of bending are seen in the microstructure. In the concave side of the curvatures, grains are compressed and assume a 90° angle to the radius of the curvature. On the opposite (convex) side they are extended (Figure 7.2).

Unfortunately, even after archaeometallurgical examination, the "jury is still out" on its source of manufacture. However, one is inclined to agree with William Fitzgerald who said that producing this fine tubing was well within the technological capabilities of native metalworkers (Fitzgerald 1990:508–509). It very well could have been a native production. Why native workers should have chosen to produce this form is unknown. However, from a mechanical engineering perspective, this design is stronger than solid "wire" of the same diameter. Irrespective of its source, perhaps the Illinois and other peoples of the Northeast chose it over strip or solid wire for bracelets, hair ornaments, and rings for this reason.

Blanks

Blank manufacture has been dealt with in detail in chapter 6, so remarks are confined here to new revelations afforded by laboratory analyses. As in the case of beads and clips, brass blanks ($n = 11$) outnumber copper blanks ($n = 8$) in the subsample. This does not necessarily indicate a *clear* preference for one or the other European-derived material type.

In terms of the artifact manufacturing process, blanks are thought to have been cut from scrap or from other larger blanks. Numerous larger pieces of metal are found "cut into," grooved, or bent at regular intervals, indicating sequential blank removal (see Figure 6.9). Compositional analysis has verified that blanks were cut sequentially from one piece of metal. Once cut, burrs on blank margins were hammered down if necessary or desired. Microstructural examination of the finished clips and beads reveals that blanks were then annealed in preparation for final forming. Nine blanks (five copper, four brass) had apparently reached this stage. One copper blank may have been worked above the recrystallization temperature (hot worked) before it was annealed.

However, it is important to note that as many blanks are left in a work-hardened condition as are left annealed. Brass blanks ($n = 7$) are left in the work-hardened state much more frequently than are copper blanks ($n = 2$). While a good deal of the localized deformation seen on these artifacts is due to shearing, additional cold hammering is noted in one or more areas of many of these artifacts. All seven brass blanks are worked heavily and differentially. Two have strain markings everywhere; another two are folded and then hammered. Of the work-hardened copper blanks, one large copper trapezoidal specimen (see Figure 6.9 b, right) was so heavily deformed that in some zones, grains are pancaked.

That so many blanks should be abandoned without having been made into finished artifacts is surprising. In one house (House 1, Area C) a single feature (F90) contains 42 such partially processed items; another (F136) has 17. In House 2, F171/172 has 20. Most are small rectangular blanks cut for manufacture of clips or beads. It is puzzling also that so many seemingly usable blanks are abandoned in a cold worked condition without ever having proceeded to the annealing stage or to being rolled into finished products. It is highly doubtful that corrosion could have been the reason. It is not likely that these artifacts would have corroded so heavily by this early date as to have lost their visual appeal even if polished.

Scrap

Copper scrap appears in this subsample in just slightly higher proportion than does brass scrap. The various ways that scrap is found to have been manipulated were covered in chapter 6. It should be restated here, however, that visually, much of the scrap appears to have undergone significant manipulation before it was abandoned. Some appears to have been extensively "worked over," exhibiting a complex work history; several of these specimens were chosen for metallographic examination precisely for this reason. Despite the scrap's worked appearance, however, metallographic analysis revealed that for just over two-thirds (n = 13) of the scrap, annealing had been the final manipulation before abandonment. Less than half of these (n = 5) had *not* been subjected to prior cold working before they were annealed. The remainder had undergone some degree of prior localized or overall cold working. Variable grain size, transgranular cracking, and lapping are the indicators of such previous activity. Seventy-seven percent of the annealed scrap is smelted copper. As pointed out earlier, copper is left annealed over four times as frequently as is brass.

The remaining five pieces of scrap (n = 5, or 26 percent) were discarded in a work-hardened state. Four of five are brass. Two show strain markings at folds and bends and two are deformed to varying degrees throughout. One has very heavy cold hammering at one end of an otherwise annealed structure. The fact that more brass scrap is discarded in a work-hardened state from these activities conforms to the pattern seen in the blanks.

Native Copper Artifacts

The native copper artifacts in this sample are fabricated using solid pieces of native copper rather than sheet. The nodules are hammered, then folded over themselves and hammered again until the finished shape is achieved. Periodic annealing restores the metal's ductility and allows for the shaping process to continue. Metallographically, the awl and the bipointed piece show clear indications of this formation process. Metallography also reveals that the point and working end of the awl were cold worked extremely heavily to harden the point, most likely in anticipation of work in the everyday world. In contrast to the sheet metal ornaments discussed above, this tool was a utilitarian implement left in a work-hardened state.

CONCLUSION

A distinctive repertoire of Illinois metalworking is revealed that demonstrates that they were working European copper and brass primarily and native copper secondarily during the mid-seventeenth century. Solid nodules of native copper were being worked into at least one finished object of known utilitarian function, while the European copper and brass are fashioned into a narrow range of finished personal adornment artifacts. European-derived copper and brass are shown to have been chosen in almost equal proportions. None of the ornaments was found to have been made of native copper.

"Raw" material of European derivation came to the Illinois as already manufactured sheet. The availability of material in this state may have been seen as a technological advantage to them, in that it saved metalworkers the step of having to reduce a metal nodule to sheet or foil thickness before proceeding. Sheet was manipulated in simple ways to form finished ornaments using techniques known by native workers for thousands of years. These processes include annealing, cold working (principally by hammering), and perhaps hot working as the primary manufacturing method. These techniques are accompanied by *secondary* techniques that included different types and degrees of bending, that is, shearing, rolling, folding, and perforating combined with some grinding, thinning, and undoubtedly some polishing. The techniques applied are not utilized in a refined manner; instead they yield rather crudely and simply formed artifacts. Most finished artifacts are left in a softened, recrystallized state, presumably because they were not meant to function in a utilitarian sphere. Significant European-derived copper-base metal is discarded in a variety of unprocessed and partially processed states. No copper-base materials were found to have been melted or cast by the Illinois.[5]

Metallographic analysis revealed interesting, but unexplained differences in the way copper and brass were worked by the Illinois. Over the entire subsample, copper artifacts are left annealed over 90 percent of the time, while almost equal proportions of brass objects are left annealed (52 percent) and cold worked (48 percent). This general pattern is repeated among the unfinished artifacts. Eighty-seven percent of the unfinished copper artifacts is left annealed, while only 37 percent of the brass is found in a recrystallized state at abandonment. Most of the 20 finished artifacts are left in an annealed state (85 percent), although more copper than brass-finished objects had been cold worked prior to the final anneal. The only finished artifacts left in a work-hardened condition are brass (n = 3).

8 / Illinois Metalworking Style in Contexts of Social Action and Technological Change

INTRODUCTION

The overall aim of this work has been to illuminate the processes, circumstances, meanings, and tempos of material transformation during the initial phases of native involvement with things European. Its focus has been on understanding the ways in which one form of European manufacture, copper-base metals, were integrated into native material culture repertoires and the role(s) they played in transforming native cultural systems as intermittent, indirect contact grew into larger-scale European presence in the mid- to late-seventeenth-century Western Great Lakes. I have argued that tracking native responses to the availability of these materials at the culture-specific and materials-specific level provides a richly textured view of early change in technological and social systems. It draws attention to how, why, and in what varied domains particular groups of native actors are found to be consuming, making, and using new materials (and things) as they adjust, negotiate, and reformulate their relations with one another and with foreigners during this crucial phase of European/native historical "conjuncture." As has been stated in chapter 1, this approach "has the potential not only to reveal unique reactions to European imports but also to make more explicit the case-specific ideas, activities, and interests that helped set larger scale processes of material and social change in motion."

This chapter draws together the evidence gathered from archaeological, documentary, and materials science sources to articulate, from a technological perspective, the responses of the protohistoric/early historic Illinois to the availability of European copper-base metal goods and materials. Findings of Illinois metalworking "style" are presented within the technological "systems" framework adopted at the outset of this study (from Kingery 1993; Figure 2.1 this volume). Technological "style" as it is used here is a multidimensional approach to artifact manufacture that integrates the technical, behavioral, ideological, social, and historical aspects of production. The technological "systems" framework views technology as an inclusive system that includes all stages and dimensions of activity and perception in making, using, even discarding things. All of these technological activities are embedded in historically specific social action, worldview, human agency, and social reproduction (after Dobres and Hoffman 1994, 1999). By integrating these perspectives, the complex ways in which the Illinois adopted new materials and absorbed them into their already changing technological, social, and symbolic systems are revealed. Further, the degree to which the integration of these new materials impacted native technologies (embedded as they are in cultural beliefs, motivations, values, and social actions) is brought to light.

After presenting Illinois copper-base metalworking style within this framework, the arenas in which and the extent to which their response to copper-base metal materials shaped Illinois social action in the earliest phases of material transformation are assessed. In doing so, the complex relations among material, technological, symbolic, and social aspects of native cultural systems as new objects and ideas are drawn into them during the earliest phases of European/Native American entanglement are made visible.

In concluding this chapter and this work, I review the important ways in which the multidimensional, materials-based approach to understanding technologies and the meaning of technological change taken here repositions the role(s) of early material acceptance within the overall trajectory of post-contact native culture transformation. I discuss how the results of this culture- and material-specific microenvironmental study of the nature, meanings and contexts of technological change in particular historic settings have brought into clearer focus the interactive relations between technology and culture. I conclude by suggesting ways in which the results of this study can continue to contribute to our inclusive understanding of native material and cultural transformation across northeastern North America.

"STYLE" IN ILLINOIS COPPER-BASE METALWORKING FROM A TECHNOLOGICAL "SYSTEMS" PERSPECTIVE

Materials Acquisition and Distribution

It should be mentioned at the outset of this discussion that given our limited archaeological understandings of the deeper prehistory of the Illinois, there is no way at present to test how extensively they used native copper and the items made from it before European smelted copper-base metals became available to them. How familiar they might have been with the refined techniques employed by other protohistoric and Mississippian metalworkers around them is also unclear. At the same time, however, this study reveals unequivocally that Illinois metalworkers were using European-derived copper-base metals primarily, while their use of native metals was secondary. As to procurement of these materials, ample historical evidence points to the availability of native copper at major ore sources approximately 575 miles from the Iliniwek Village. These sources had been "mined" for thousands of years by native peoples of the midcontinent and were undoubtedly familiar to at least some peoples residing outside a direct exploitation sphere. Nuggets of float copper deposited in glacial drift were also procurable over a wide range south of the Lake Superior region. The Iliniwek Village is within this distribution area (Rapp, Allert, Vitali, Jing, and Henrickson 2000:11–12) in which nuggets could have simply been picked up (Pease and Werner 1934:347). While it makes sense that the Lake Superior region is the source of the two native copper specimens from the Iliniwek Village, it cannot be stated with certainty based on the results of the compositional studies.

There are several potential sources, both foreign and native, from which the Illinois could have obtained European copper-base metal materials. While unauthorized *coureurs de bois* were probably plying the rivers from time to time with wares for exchange, these possible visits remain unrecorded (Blair 1996:1:228 f 164). On the other hand, there is clear documentary evidence that the Illinois were getting materials directly from the French at missions in Wisconsin (Thwaites 1896–1901:51:49). The Illinois are known to have been inhabiting multiethnic villages near the mission at Green Bay during the late 1660s in order to gain greater, uninterrupted access to the goods at the mission and to connect themselves with native peoples who had direct contact with the French in eastern trading centers (Blair 1996:1:321–322). They are also known to have visited the straits of Mackinac in the 1670s to trade (Blair

1996:1:350). These kinds of early actions on the part of the Illinois have led Raymond Hauser (1973:3) to argue that the Illinois "very early became closely identified with the Europeans, rapidly forming an economic and political allegiance with French traders and missionaries."

Their early trading activities served multiple purposes. The important lines Father Claude Dablon (Thwaites 1896–1901:54:167) writes about the Illinois coming to St. Esprit, probably from the Iliniwek Village, illuminate their motives. He relates:

> The Ilinois, tribes extending toward the South, have five large Villages, of which one has a stretch of three leagues, the cabins being placed lengthwise. They number nearly two thousand souls, and repair to this place from time to time in great numbers *as Merchants,* to carry away hatchets and kettles, guns, and other articles that they need [emphasis mine].

Clearly, the Illinois were trading for items they themselves intended to consume, but researchers have also taken this passage to indicate that they then traded this material out to other tribes farther to the south and west. In this way, they became "middlemen"[1] in the trade to the interior (Blasingham 1956:195–196; Brown and Sasso 2001:20). At the same time, they were becoming more and more visible to the French, who were clearly using the trading centers and the lure of European goods as an opportunity to present the Christian faith to those who came to trade. Their desire to be visited and proselytized follows directly in Dablon's account. He (Thwaites 1896–1901: 54:167) goes on to say:

> During the sojourn that they make here, we take the opportunity to sow in their hearts the first seeds of the Gospel. Further mention will be hereafter made of these peoples, and of the desire which they manifest to have one of our Fathers among them to instruct them; and also of the plan formed by Father Marquette to go thither next Autumn.

What particular resources the Illinois were trading is known in a general sense. The documents indicate that it included (but may not have been limited to) slaves, animal pelts/skins, and bison robes (J. Brown 1961:69; Hauser 1973:82–88; Thwaites 1896–1901:59:67, 175, 177). Although beaver pelts were thought to have been the main resource that brought native peoples trade goods, the Illinois were not traditionally thought to have been good beaver hunters. According to Deliette they did not "esteem the beaver" as highly as the Miami (Pease and Werner 1934:393). However, slave taking and trading among them is documented more than once by Marquette, who specifically mentions that the Ottawa were involved in receiving the slaves and that target tribes "to the south and west" were the unwitting givers. According to Marquette (Thwaites 1896–1901:54:91, 59:127), these tribes had no knowledge of the copper and iron that the Illinois were receiving for slaves.

It is not known whether the Illinois became the native source from which trade commodities reached these far-flung groups during this early period. However, Deliette records that by the end of the century, "several" groups from the Missouri River area came "often" to trade with the Illinois. These included the Osage and the Missouri, who received "hatchets, knives, and awls, and other necessary things" from them. While these groups had formerly been at war with the Illinois, they maintained peaceful relations to keep open access to trade goods (Pease and Werner 1934:387–389).

Early on in their encounters with the French, the Illinois could very well have received copper-base metals as part of the intercultural diplomacy that framed native/European communications.[2] As mentioned earlier, presentation of gifts (in trade goods) rapidly became an indispensable way to bestow prestige, recognize rank or status, mark an auspicious occasion, negotiate and cement a new social relationship and/or alliance, and create new obligations (Washburn 1967:50; White 1991). Mutual gift exchange was an important, no doubt expected, component of Marquette and Jolliet's interaction with the Illinois upon their arrival at Peouarea in 1673 (Thwaites 1896–1901:59:119–122). In another example of these types of ritualized situations, Claude Charles Le Roy de La Potherie (Blair 1996:1:330–331) relates Nicolas Perrot's meeting with the residents of a multiethnic village (which included some 15 "cabins" of Illinois) near Green Bay. Perrot gave the young male attendees his gun and the elderly males his kettle. In a florid, yet quite revealing speech that accompanied the gift giving, Perrot extols the dawn of a new era of native prosperity as a result of their alliance with the French and their acceptance of technologically superior French merchandise. He beseeched the older men:

> To you who are old men I leave my kettle; I carry it everywhere without fear of breaking it. You will cook in it the meat that your young men bring from the chase, and the food which you offer to the Frenchmen who come to visit you.

He distributed metal awls, knives, and glass beads to the women saying,

> Throw aside your bone bodkins; these French awls will be much easier to use. These knives will be more useful to you in killing beavers and in cutting your meat than are the pieces of stone that you use. . . . See; these [beads] will better adorn your children and girls than do their usual ornaments.

Later, in 1680, Recollet Father Louis Hennepin provides direct evidence for these exchanges with the Illinois when he reports that LaSalle's expeditionary party gave presents to them in exchange for information and to affirm treaties between the Miami and the Illinois. LaSalle had given Hennepin

> . . . ten Knives, twelve Shooe-maker's auls or Bodkins; a small Roll of Tobacco from *Martinico,* about two Pounds of *Rassade;* that is to say, Little Pearls or Rings of colour'd Glass, wherewith the Savages make Bracelets, and other Works, and a small Parcel of Needles to give to the Savages . . .

for "his own use" in his dealings with them (Thwaites 1903:180).

From an inter-regional perspective, native/native trade systems are well known to have been major avenues through which native people of the interior originally received and exchanged trade goods (Drooker 1996a, b; Drooker and Cowan 2001). Such networks were particularly active on the frontier edges of the trade and in early periods before Europeans arrived and shifted at least some native attention to trading centers. Many networks were certainly in operation in late protohistory (Drooker and Cowan 2001; Stothers 2000). At least one of these trade routes may have been through the Lower, rather than the Upper Great Lakes. David Stothers (2000:70) argues that an Algonquian trade axis extended into the interior from the "middle St. Lawrence River Valley, through the lower Great Lakes, to the western end of Lake Erie." He goes on to say that this route had been known to Champlain by 1603 (Stothers 2000:70). Whether this trade route extended westward into the Mississippi Valley is unknown. The suggestion that the Illinois may have migrated into the Mississippi Valley around 1640 from somewhere below Lake Erie where protohistoric site locations and a narrow range of European-derived copper-base metal forms are documented places them in a strategic position to have known of these routes and these sources of material, even to have obtained some before their move (Ehrhardt 2004; see Stothers 2000 for description of these artifacts). However, what role the Illinois may have played in drawing trade goods from the Ohio Valley into the Illinois Country once they moved there is as yet unclear.

Although there is no *direct* evidence for it at this point, this begs consideration of the possibility that the Illinois received copper-base material from other-than-French sources. The geographic location of the Iliniwek Village in the lower reaches of the Upper Mississippi Valley does not preclude native/native access to goods coming from the Northeast, the mid-Atlantic or the interior Southeast. However, the extent to which this actually took place is undocumented historically and cannot be demonstrated reliably by the Iliniwek Village trade goods assemblage. Drooker and Cowan (2001:103–104) have suggested that by the early seventeenth century, early types of trade goods (pre-1625) had reached protohistoric Fort Ancient peoples from the Central Mississippi Valley and from as far north as southern Ontario. Drooker (1996:174–175) has documented the links between Madisonville (Ohio River Valley) and Central Mississippi Valley/Southeast based on the appearance there of a single Clarksdale bell.

Marvin Smith (1987:52) maintains that the Spanish did not trade kettles during the early historic period Southeast. The lack of kettle parts, scrap, and unfinished material on the sites he studied led him to conclude that tinkling cones and other copper-base metal ornaments were made for the interior southeastern trade by "European entrepreneurs" (Smith 1987:37). While clip forms and sizes from the Iliniwek Village are very much congruent with those found on at least one protohistoric site in Alabama (Smith 1987:47), trade in the objects themselves is unlikely; there is too much evidence from the Iliniwek Village that the Illinois were making them locally.

Unfortunately, the lack of sites in the zone between the Lake Erie region, Fort Ancient, and the Iliniwek Village precludes tracing migration and/or trade routes. Compositional studies aimed at comparing the "fingerprints" of Lake Erie and Madisonville copper-base trade metals and those from the Iliniwek Village may well shed some light on potential connections.

What form copper-base metal was in when it arrived at the Iliniwek Village is not clear. We see few kettle parts and no spent or broken kettles. This could mean one or a number of the following: 1) the trade kettle parts or spent kettles are disposed of in areas of the site not yet investigated, 2) every bit of the kettle was cut into small pieces leaving only the bails, lugs, and iron support bands for recovery by archaeologists, 3) that the copper-base metal that the Illinois received came in already reduced form. All of the above are possibilities; however, since no large pieces of kettle bodies are found that would represent the initial stages of kettle reduction, the second and third options are more likely explanations. The single lug in the industry, around which virtually all body material has been cut away, attests to optimized removal of whatever sheet metal was available.

Once copper-base metal material arrived at the Iliniwek Village, how it was distributed among the people who lived there is also uncertain. Although it does not appear in every pit feature originally selected for this sample (20 features in the sample did not contain any copper-base metal), it is widely distributed across all areas of the site excavated to date. The pattern of its distribution is congruent with the pattern of pit feature distribution across the site. There is also significant copper-base metal in the occupation levels above the pit features.

From an ethnohistorical perspective, it appears that Illinois chiefs and/or other specially ordained figures were involved in or controlled at least some aspects of procurement or distribution of trade goods. While the Jesuits and the travel documentaries do mention that special envoys or tribal representatives were often the recipients and the donors of gifts during ritualized communications, the *Relations* also point to specially designated persons who were esteemed among the Illinois for being involved in trade with Europeans (Thwaites 1896–1901:59:167). Margaret Brown (1979:255) suggests that in its earliest phases "higher ranking males dealt in the fur trade," and that it brought "new opportunities for advancement for men from less prestigious families and younger males who could attain wealth from both furs and slaves captured in warfare." Prestige may have come from what Mary Helms (1993:101) has described as the

> . . . highly regarded skills and personal characteristics of the acquirer, who has to deal in some fashion with a conceptually distinctive foreign realm qualitatively defined as involving the sanctified, the mystical, or the power-filled.

Numerous early references attest to the notion that native Algonquian and Siouan peoples of the interior attributed magical, spiritual, even "control" power to the French and foreign technology (Blair 1996:1:158–163, 309, 311; White 1994:380).

Chiefly control over the allocation of the materials once they reached a native destination appears to be absent among the Illinois, but this cannot be stated with certainty. From an economic viewpoint, Marquette remarks that upon returning to their own villages after obtaining European goods, Illinois chiefs/middlemen were canny traders "[giving] hardly any more than the French" in exchange for buffalo robes (Thwaites 1896–1901:59:175). How control over the subsequent circulation of finished or unfinished material among villagers actually worked is unknown. This pattern of distribution may stand in stark contrast to that which was known for the early-seventeenth-century Powhatan confederacy, in which Powhatan bought all the available English copper himself and used it for his own purposes. Recipients were primarily other *weroances*[3] (Roundtree 1989:55).

Design

Except for the native copper artifacts, all of the reworked materials in the Iliniwek Village copper-base metal sample appear to have been designed for use as ornamentation or personal display on the person, on clothing, or on accessories. None of the finished artifact types featured in the Iliniwek copper-base metal industry are unique to the Illinois nor are they unique in the midcontinent. Bead forms are reported from prehistoric times, and tinkling cones, spirals, coils, and triangular pendants made of reworked metal are also found earlier elsewhere in the Western or Eastern Great Lakes. Clips and triangular pendants appear to be rare for their time in the Western Great Lakes. Although they are known farther to the south, clips of the size and staplelike form found at Iliniwek are not common in the midcontinent either before or after this time. Pendants emerge more commonly during the later decades of the seventeenth and early eighteenth centuries. The particular "open midsection" design features of some of the Iliniwek Village tinkling cones are unexplained and may be unique. The design may reflect a local adaptation of a panregional ornament style and as such may signal some specialized use. However, further investigations of additional inventories are required to verify this.

Some of the artifacts the Illinois fashioned had been, and sometimes continued to be, worked in other media among native peoples. Numerous authors have argued that trade metal simply became yet another medium through which native forms were expressed. For instance, tinkling cones are widely thought to have been modeled after jinglers made of deer dewclaws. Perforated triangular pendants resemble catlinite pendants of the same form. Cylindrical metal beads were similar to shell or bone forms, which both preceded and were concomitant with them. Other metal forms, like the clips or the fine hollow tubing from which coils, rings, and bracelets were shaped, appear to have had no formal or functional antecedent. Of all of these precursor forms, only a few shell beads are seen in the Iliniwek Village assemblage.

Most enigmatic is the single-lobed and double-lobed tubing. The source of these materials remains uncertain. As has been argued, however, achieving its design is well within the technological capabilities of Illinois and other indigenous copper-base metalworkers. Why they should have elected to carry out this particular design is also unknown. It does appear to be chosen for use in manufacture of ornamental objects requiring added strength if used in daily life, however.

Manufacturing

At the Iliniwek Village, nodules of native copper were worked into solid, finished artifacts using fundamental techniques native copper workers in native North America had known for well over 6,000 years. Repeated sequences of annealing and heavy cold hammering were the primary shaping operations through which the Illinois produced the single utilitarian artifact in the sample. Another object, whose function is not known, also shows signs of this same kind of manipulation. It, however, was abandoned in an annealed rather than work hardened state.

The Illinois strategy for reworking European-introduced copper and brass kettle sheet into items of personal adornment was simple and straightforward. Kettles or kettle body fragments were first cut apart using several possible types of shearing actions. The overall thinness of the metal undoubtedly simplified this process. Indigenous tools and European implements, including iron knife blades and scissors, were likely called on for the purpose. Fragments were then cut into desired sizes and shapes such as rectangles, trapezoids, and long strips to be used as blanks or preforms for finished objects. Burrs were sometimes, but not routinely, ground or hammered down. Unfinished artifacts are recovered in various states of manufacture. Partially processed blanks and tubing are found in cut, bent, rolled, and/or folded condition.

Not all available metal was cut into recognizable blank or preform shapes; perfectly usable cut-up material was abandoned, lost, or rejected as "scrap" before ever being used. Before abandonment, scrap is found having been reworked through various, often multiple, processes of hammering, bending, folding, rolling, grinding, perforating, and annealing. By the same token, many blanks were prepared to various degrees through cold working (hammering) and annealing, but were never formed into finished artifacts, having been lost or discarded (in some cases in significant quantities) and ending up in refuse pits instead. Scrap, blanks, and partially finished artifacts appear to have been discarded for no apparent reason.

To make finished objects, blanks were most often annealed before final shaping. Shaping techniques were unsophisticated. Artifacts were formed by bending, folding, or hand rolling blanks or tubing around a mandrel or form. In some cases, shaping also appears to have been achieved without the aid of a form. In a few instances, especially in tinkling cone and bead manufacture, pressure was applied to the object to steady or secure it while other operations, such as shearing, shaping, or fastening, were being performed. Perforations were accomplished by first thinning the metal, then carefully punching a rounded, sharpened implement through it. Most finished artifacts were left in an annealed condition. A few are given a small amount of cold working afterwards.

The Illinois appear to have had no preference for brass or copper in the production of finished objects, although the manner in which these materials are worked differs. Among the unfinished artifacts, copper is most often left in an annealed condition, whereas brass is found in some state of work hardening four times as often as copper. All of the finished artifacts that were left cold worked are brass. Why the Illinois should be differentially working these materials is unexplained. While brass becomes harder and stronger than does copper upon cold working, why this should be differentially desired is unclear.

Looking at native copper-base metal object manufacturing processes over the *longue durée,* important continuities in the primary ways that native peoples worked this material extend through the protohistoric/early historic period. Shaping the metal by cold working, annealing, and hot working still prevailed among native craft workers. Although the Illinois melted and reshaped glass trade beads, no intentional melting, smelting, or casting of native or European-derived copper-base metal emerged during this time (M. Brown 1979:77–78). Sheet and foil stock remained the main form of raw material manipulated, except the stock now came to them already prepared. The production of sheet or foil from copper nodules, a critical, time- and energy-consuming step in the prehistoric manufacturing process, was now eliminated with the availability of already reduced and thinned sheet from metal kettles.

However, comparing Illinois methods to those employed earlier in prehistory, we also find some fundamental discontinuities. There is no indication that the Illinois made or used sheets in the elaborate, complex manner known in other past or contemporary major North American metalworking traditions of the midcontinent and Southeast. They do not appear to have hammered nodules of native copper into thin sheets or into thin foils for layering into or onto finished objects, for sheathing objects manufactured in other media, or for fashioning intricately designed and executed ceremonial objects. They do not rivet pieces of it together mechanically to make larger sheets. Although kettle sheet metal of regular thickness was available to them (albeit in potentially fragmentary form and in limited sized pieces), they do not appear to have made common practice of reducing it in thickness any further than when it was received. There is no evidence for the use of joining, sinking, embossing, riveting, or repoussé techniques, or for the manufacture of or trade in elaborate, decorative finished objects.

With the arrival of smelted metal then, *primary* native manufacturing processes still continued to focus on the use of annealing, cold working, and to a much lesser extent hot forging methods; however, these processes are now followed by much less elaborate *secondary* techniques of tooling and bending. The sophisticated ways in which the Illinois' prehistoric precursors created heavy artifacts for use as tools or turned fine foils into magnificent status and power emblems seem to have given way to making very different, less complex, but still ornamental, artifact forms. For the protohistoric Illinois, this meant shearing, working, annealing, and forming simple, unelaborated decorative objects expediently and with apparently little concern for the economical use of preforms or of all of the available sheet.

The amount of scrap found in the sample, the number of artifacts abandoned in various stages of manufacture, even the seemingly careless and hasty, yet redundant, way some of the artifacts are fashioned argue strongly that a significant number of the finished copper-base metal artifacts found at Iliniwek were made locally. This is not to say that the Illinois were not sophisticated or experienced metalworkers; the techniques that allowed joins to meet neatly and overlap tightly and edges to be smoothed were probably well known to them but apparently not frequently employed. Perhaps for some unknown reason, expediency was more important than fine finishing technique. A more important question might be whether these artifacts were actually "satisfactory" *to their makers* in a technical, symbolic, or other sense (after Latta and Anselmi 1995:4).

Use

Bent or rolled sheet metal was formed into simple ornaments that were meant primarily to be 1) used singly or with other artifacts of the same or different media, and 2) affixed to body, and/or to the hair, clothing, receptacles, or accessories either through stringing or by direct attachment. It is apparent that in contrast to single large, symbol badges, plaques, and gorgets displaying well-recognized iconographic messages of power, prestige, and/or spiritual connection, these were small decorative objects that served as single design elements used to create an overall decorative, yet potentially equally powerful aesthetic and symbolic display. Some of these items, including small bead and clip ornaments, were likely used in multiples, having been applied to fabrics or backing one or several at a time, and arranged in particular patterns that formed an overall design or design style. It is not difficult to imagine that individual metal beads and clips of various sizes and colors (red/copper, yellow/brass) were chosen in much the same manner as a beadmaker might select a particular size, shape, and color of glass or shell bead in order to compose an intricate or meaningful design. While the metal beads in the Iliniwek industry still attached to the rawhide thongs offer only meager testimony to such practice, the numerous examples of the purposeful placement of beads and clips on material backings elsewhere in the Western Great Lakes attest more strongly to it. Tinkling cones may also have been used in this manner as well.

Objects of other organic and inorganic media, including shell and glass beads, leather, animal parts such as claws, feathers, and bones, colored stones, clay, and porcupine quills were likely incorporated into the design process and into finished products. The materials to which these materials were attached may have been prepared in a particular way then painted or dyed, incorporating additional color and textural enhancement to the finished pieces. Among the Illinois, animal hides were colored red, black, and yellow (see Pease and Werner 1934:339). Marquette (Thwaites 1896–1901:59:127) comments on the skill with which the Illinois fashioned the red "Scarfs" worn by the "Captains." These were made of the "hair" of bears and wild cattle undoubtedly dyed the red color to which he refers. Even the color of the metal itself is known to have been set off or embellished or charged symbolically by the use of ochre.

While ample evidence has been presented to demonstrate that the Illinois fabricated copper-base metal ornaments at Peouarea, there is little *direct* or *specific* information as to how and in what contexts they were actually used as bodily or clothing adornment. Copper-base items, mostly coils and tubes, are found in mortuary contexts deposited primarily with children (adults = one adult male, children = four, females = none) (M. Brown 1975:39–41). These may well have been ornamental accouterments as opposed to burial offerings, but Margaret Brown is not specific as to their placement in the graves. However, coils of double-lobed tubing appearing in burial association with children and infants indicate that in these cases, bodily adornment was a feature of ascribed rather than achieved status.

Based on the evidence gleaned from contemporary ethnohistoric descriptions of Western Great Lakes peoples and the distribution of these artifacts in the archaeological record, it is quite likely that the Illinois used these various forms of adornment in life as well as in death to visually enhance their bodies or to embellish their clothing and accessories. Among their many potential use contexts, they were likely utilized alone or more probably in combination with other elements as hair ornaments, ear bobs, necklace parts, jewelry, and as fringe ornaments on clothing or as design elements on accessories such as belts, scarves, leggings, garters, bags, pouches, knife sheaths, or decorated panels (Bray 1978:56). Certain accessories were undoubtedly embellished with designs, motifs, and other imagery representing life forces or spirit guardians. In this manner the wearer was connected in a symbolic way to the spirit world, and that connection was communicated visually to others.

While there is no reason to suspect that ornamented clothing, accouterments, and jewelry items were always reserved for use (or display) on special occasions, some of these items may well have been brought out for context-specific uses or as part of special ritual regalia or paraphernalia associated with particular types of events. One such moment is described by Deliette, who recounts seeing a prospective Illinois bride leading her family's procession to present gifts to her groom. The bride, he observed, "march[ed] ahead well adorned with shoulder straps, glass beads, porcelain, and bells, so that one who heard them marching would think they were mules" (Pease and Werner 1934:333). In a somewhat different context, Nicolas Perrot claims that Ottawa men were "decked with the most precious ornaments" that they possessed when they gave feasts preceding wars (Blair 1996:1:54). Although this cannot be known with certainty, this was likely true among the Illinois as well.

What appears clear is that regardless of how they were used, these objects served as shiny, colorful ornamentation that enhanced both the sound and reflective qualities of the metals and the appearance of the wearer. Light shining upon these metals from the sun (considered a light- and life-giving deity among Central Algonquians, including the Illinois [Blair 1996:1:48, 59; Thwaites 1896–1901:54:187]), augmented the overall visual aesthetic of sound, color, reflection, and movement (see Dubin 1999:24). While unsubstantiated, it might even be suggested that the sun warmed the metal, potentially reinforcing copper's meaning as a life-sustaining force (see Herbert 1984:280). These qualities probably made articles of European derivation very desirable (M. Brown 1979:262). While Margaret Brown (1979) has argued that these artifacts may have been differentially utilized by higher status individuals, the wide distribution of these materials from Iliniwek does not bear this out to date. Whether they were material expressions of elevated ascribed or achieved community status, gender or age differentiation can only be suggested because little evidence is available from the archaeological record. To the best of current knowledge, no specific mention is made in the documentary sources of the time.

The degree to which consumption and use of reworked copper-base ornaments among the Illinois actually impacted existing social interrelationships within their community in this early time is also unknown. One insightful study (Cannon 1991) illuminates the changing relations among gender status, material demand, and material expression within and across Huron gender divisions during the late protohistoric/early historic period. Using adornment as a means of competitive display, Huron women demanded greater and greater amounts of adornment items from potential suitors; these suitors provided it depending on their success in trade, warfare, and theft. Since men's status rested on how well they accomplished these deeds, men were encouraged to participate to an ever-greater extent in these activities to prove themselves worthy. According to Aubrey Cannon (1991:147), display through imported media took on new importance as women's influence over men's external activities grew. It is one way in which the elaboration of adornment items on Huron sites of the period can be explained.

Perception

Here, perception is taken to refer to the "meanings" the Illinois attached to copper-base metal materials and the objects made from them. Much has been written about the symbolic, aesthetic, and economic "value" placed on copper as a raw material by native peoples of the Northeast (see chapter 4). Archaeologists have hypothesized that as copper-base trade metal became more widely available to native peoples, the use of native copper declined sharply as did notions about its symbolic worth (Martin 1999:180). Susan Martin (1999:183) has said, "the significance of copper depends upon the community and culture in which one lives." Regrettably, we know little about the particular symbolic value the Illinois placed either on copper (native or European-derived) as a raw material or on the ornaments made from it. The documents reveal that like the Sioux, the Illinois believed (at least at first) that the French were spirits and their words came from "the other world" (Thwaites 1896–1901:55:215–217). Bruce White (1994:369) has suggested that the Ojibwa and Dakota of the Western Great Lakes called the French "spirits" because they appreciated French technology and considered it "powerful." Cornelius Jaenen (1974:265) had earlier recognized (as cited in chapter 4) native appreciation for European technology, especially for knives, hatchets, kettles, beads, cloth, and firearms as items of "aesthetic, magical, or purely decorative and fascinating worth not their economic value."

While this may well have been the case in large part, nothing in the contemporary documents reviewed supports a belief specific to the Illinois that European copper itself or copper-base metal kettles had "spirit" or came from "the other world." To the extent that this may be true for them, we may have to fall back on George Hamell's (1983, 1986) and Christopher Miller and Hamell's (1986) thoroughgoing and well-articulated treatments of the shared "metaphysical" meanings of copper to native Algonquian, Siouan, and Iroquoian peoples of the Northeast. It will be remembered that Hamell (1983, 1986) assigned this material to the underwater world and to the Underwater World Grandfathers, and described it as "symbolically charged," having life-affirming and restorative qualities. Further, for peoples of the Northeastern Woodlands, the color red (the color commonly associated with copper) mediated between light (life) and darkness (death) and was suggestive of high emotional states and of blood. This may well be true for the Illinois, but no connections between the color red and copper as a material are made in the documents. Only the associations with war, blood, and the color red are confirmed for the Illinois (Pease and Werner 1934:384). The extent to which copper (or brass) might have fit into this system of symbolic and ideological meaning is not known.

Jaenen's (1974:276) remarks on the material structure of interaction put us on much firmer ground with reference to the Illinois systems of meaning attached to the acquisition of copper-base metal material goods. He states, "this exchange, for the Amerindians, had a symbolic or diplomatic meaning and was in reality viewed as an exchange of gifts which established rank and prestige." The Illinois participated in innumerable ritualized visitations and communications with other native groups or with the French that involved the exchange of trade goods. These materials were recognized by all parties as essential to the success of the encounters and as having special prestige and high exchange value (see endnote 2 this chapter). Kettles were often significant material elements in these exchanges. Nothing is mentioned, however, of finished ornaments, kettle parts, sheet metal, or tubing being material features of these ritualized expressions of good will/obligation.

Despite the strong likelihood that Jaenen is correct about the symbolic and prestige value that copper-base metal artifacts had in Illinois exchange contexts, these objects likewise quickly assumed significant economic value for them. From the late 1660s, kettles (or quite likely fragments of kettles) are known to have been used by them as items of trade to interior tribes. In this sense, trade relations the Illinois engaged in with both French and native intermediaries were indeed economically motivated to no small degree. They would escalate as the historic period intensified and the Illinois became better established and more influential in trade to the interior.

While no whole kettles are found early on at the Iliniwek Village, it is beyond question that by the end of the seventeenth century, intact copper-base metal kettles became well integrated within the Illinois social system as important material symbols in ritualized gift giving. They are given in offering to shamans or "medicine men" for healing of the sick; a kettle painted red accompanied by a hatchet prepared in the same manner meant death for a captive (Pease and Werner 1934:384). Kettles also figure prominently in mortuary rituals. While it is not known whether the Illinois routinely deposited complete or ritually "killed" kettles in mortuary contexts, Deliette describes mortuary ceremonies in which kettles are indeed placed in graves (Pease and Werner 1934:357, 366). They also become part of the offerings that "clothe" women whose husbands have died. The amount of these and other European materials presented was thought to show the esteem in which the deceased was held. There seems to be no preference for brass or copper kettles in these routinized exchanges. According to Deliette (Pease and Werner 1934:357), "If he [a donor] has given a yellow kettle, he receives a red one; if he has given a small kettle, he receives a hatchet." Kettles in the homes of men who remarry outside the previous wife's family are also ritually broken in protest by the female relatives of the dead woman (Pease and Werner 1934:361).

Copper-base metal kettles also became prominent features of bride price, given as items of ritual exchange with other trade goods by the prospective groom "according to his wealth and the esteem in which the girl is held" (Pease and Werner 1934:331). Gifts of wealth in trade goods or "merchandise" such as guns, cloth, and kettles are emphasized. Items may be turned back and forth several times, always augmented with more offerings, before settlement is reached. This is followed by additional gift giving (Pease and Werner 1934:332–333).

Reuse and Discard Technologies

What must be recalled here is that Illinois acceptance and reworking of copper-base sheet metal from kettles manufactured in Europe into ornaments clearly constitute a "reuse" technology in and of itself. Within this context, the mechanics of kettle sheet metal reuse among the Illinois have already been discussed in significant detail. In addition, the technological reasons why it is suspected that some of the artifacts, specifically the clips, may have been removed from one use context in anticipation of use in yet another have also been presented.

The amount of copper-base metal material discarded or abandoned in refuse pits at the Iliniwek Village is puzzling considering the emphasis placed by the Illinois on its decorative uses and on its purported economic, symbolic, and aesthetic worth. It remains difficult to sustain arguments about copper's high value to the Illinois, when so many prepared blanks and "scrap" of usable size and condition went unused and/or are discarded. Unlike nuggets of native copper, which were curated by native peoples as items possessing "power" and "magic," these discarded artifacts appear not to have been revered in this way (Halsey 1992:1). Howard Winters (1968:184) has argued that items of "special value" are assessed as valuable because they are made of a rare, often imported material (as copper); they are usually ornaments or other items of ceremonial regalia and they appear in special, segregated contexts. While some of these conditions are met here, the fact that so many Iliniwek copper-base metal items and materials are found in the trash does not meet the "special, segregated contexts" condition set by Winters. Perhaps its symbolic and/or ritual significance had been "lost" in some way, or was situational.

One possible answer is that the supply of copper-base trade metal to the Illinois was already so regular and reliable at this early time that this material had already lost whatever special status it once held among native peoples (Bradley 1987:186). While this explanation might address satisfactorily the symbolic aspects of its value, it does not take into consideration that the Illinois were thought to be trading the material into the interior. In these regions, copper-base and ferrous materials were still quite rare and were considered novelties. As such, it may well have still retained significant symbolic and social value. In this way, it would have remained a valuable economic means through which the Illinois could have strengthened their foothold as purveyors of these goods and heightened their influence in interactions with neighboring groups.

ARENAS AND CONTEXTS OF TECHNOLOGICAL CHANGE

It is clear that early Illinois technological response to new copper-base metal materials was multilayered. At this point in their contact history, acceptance of this "new" product into their material culture repertoire represents an overall additive rather than replacement component to an intact material and technological system (after Rogers 1990:105–108). Conversion of these newly acquired materials into new and already existing forms of personal adornment from objects originally manufactured by foreigners for a completely different use reflects the Illinois' early and widespread acceptance and integration of these *materials,* not necessarily these *objects,* into their own sociotechnological systems. It further shows that even before their first encounters with Europeans, assimilation of these newly available materials was *not* need driven in a technological or a use-context sense, as researchers have come to realize. It is very likely that both the French and the Illinois knew that the objects Europeans offered were "superior" in a technological way to those which the Illinois traditionally fashioned. However, the varied creative ways in which the Illinois used the otherwise utilitarian copper-base cooking utensils they received do not accord fully with such an interpretation. As this research has also brought to the fore, there are several likely reasons the Illinois may have "needed" (sought?) these products early on. One, of course, was for the raw materials with which to make objects of personal adornment. Another was for exchange or perhaps for prestige or advantage in gift-giving and alliance-forging activities with each other and with their neighbors. Another was for trade with groups farther into the interior who were desirous of foreign products but had no contact with Europeans. Yet another was because it may well have been recognized as a material of great spiritual power and symbolic worth, offering protection, good health, and good fortune. These reasons notwithstanding, there is ample evidence that continued desire for these products set in motion larger scale internal processes of material and social transformation that intensified as conjunctures with Europeans continued in the historic period.

Over the course of this research, it has been demonstrated that from their initial introduction to French copper-base metal artifacts, changes and continuity are seen at many points in the structure of the Illinois technological systems. These processes range from the strategies connected with procuring these materials, through the practices involved in and the uses to which newly fashioned articles are put, to the potential shifts in the ways these materials were perceived and even disposed of. Local decisions to participate actively in exchange and social relations with Europeans and to encourage foreign presence among them were important stimuli for their ever-intensifying entanglements; these involvements clearly impacted the timing and tempos of these transformative processes. These choices surely had immediate and long-range consequences for the trajectories of Illinois culture change. Hauser (1973) has argued that they sped up the processes of depopulation and cultural disintegration.

At the same time, these early decisions and activities also resonated strongly in many other aspects of Illinois life. Hunting and processing patterns were altered and intensified as the demand for native resources grew. Conditions and frequency of warfare were rearranged as alliances were realigned, often prompted and/or mitigated by French interests. Traditional belief systems were challenged. Enemies became friends (or at least tolerable neighbors) in order to gain access to trade goods, as Illinois peace accords with the Sioux in the late 1660s demonstrate. While it is not known for certain, slave taking and trading may well have intensified. All of these kinds of activities took place in theaters of social action in which the Illinois were continually repositioning themselves. Special interests of individuals and/or entities within the larger unit may have engendered new social and political roles that required adjustment of the group structure itself. Thus, while cultural impact has been termed "minimal" in the initial phases of these processes, these findings illuminate the true complexity of cultural dynamics, even in this early period (Emerson and Brown 1992:78).

Importantly for this work, the materials-based examination of copper-base metal artifacts and materials from the Iliniwek Village has elucidated the culture-specific technological processes, practices, perceptions, and sequences involved in turning copper-base metals into desired products with a "native" identity rather than a European one (Bradley 1987:168). The techniques used by the Illinois did not reach the level achieved by postcontact Onondaga metalworkers who reworked these same materials with greater refinement, even casting them (Bradley 1987:176–178). At the same time, however, this investigation has provided an important opportunity from which to assess change over the *longue durée* in metalworking "styles." James Bradley's evidence for native casting copper-base metal during the historic period among the Onondaga notwithstanding, is there one persistent, traditional "native style" of copper-base metalworking in native northeastern North America?

The basic, unchanging way in which copper was worked and shaped in the native North American midcontinent through prehistory and early contact was through a *primary* technique of annealing and cold hammering processes with some hot forging. Ursula Franklin and colleagues (1981:38) have called this phenomenon a "single technological pattern . . . [that] persisted through time and cross-cut traditionally recognized cultural and linguistic boundaries." Within this primary forming process, various metalworking traditions arose that utilized a range of combinations of *secondary* working or *tooling* techniques (Franklin et al. 1981), which I argue here gave them a distinctive "stylistic" character. When all dimensions of copper perception and use, which Heather Lechtman (1977, 1994) claims are essential to defining metalworking style, are taken into consideration, one could not conclude that only one metalworking style prevailed through prehistory. Throughout the *longue durée* of its use, native copper and smelted European copper and brass artifacts have taken on different form, exhibited varying sophistication of manufacture, appear in diverse contexts, and were meant to convey distinctive iconographic or symbolic meanings. These differences serve to distinguish "styles" in time and space. Artifact types such as Hopewellian "yo-yo" type ear spools are technically quite complex. Others, like the Spiro plates and badges, were not only technically and artistically sophisticated but also were powerful and effective material symbols of chiefly prestige.

While native northeastern North American copper-base metalworking practices never developed past hammering and annealing into melting, smelting ores, casting (until the historic Onondaga example cited above), and alloying, significant variation is seen in the ways in which native metalworking practitioners, at times highly skilled specialists, interpreted technologically the metal they handled. Paul Craddock (1995:122) maintains that North American metallurgical practice "was the true summit of achievement in the fashioning of native metals." Equally significant are the diverse, creative ways that it was interpreted culturally by the peoples who used it.

CONCLUSIONS

Set in a critical yet poorly understood time of unparalleled upheaval in the history of North American indigenous populations, this project has brought into sharper focus the imminently transformative relations between European goods and materials and native responses to their availability. Having critiqued and put aside (as have others as well) those acculturation models in which native peoples are seen as passive and powerless in the face of a dominant invading culture with its virtual arsenal of technologically "superior" merchandise, this project revisits these questions from a technological perspective. First and foremost, in this research, technology has been resituated as a variable rather than as a determinant in investigating both micro- and macroscale processes of native material and social transformation. Its role as a force of change in these systems has been reexamined not as a prime mover but as one of many vectors that exerted pressures on native cultures from their earliest encounters with foreign influence and interests. This has been accomplished by adopting a broadened definition of technology as a "system." This approach takes into consideration the multiple dimensions and full range of contexts of materials acceptance and use from procurement to discard. I have focused on a culture-specific case of acceptance and use of one particular material, copper-base metal, concentrating on bringing to light the metalworking "style" of the protohistoric Illinois for the purpose of bringing to light their responses to the material. In this way, I identify, isolate, and elucidate those microenvironments of change that are found to resonate in and provide the impetus for larger-scale processes of material and cultural transformation in situations of culture contact.

At the same time and very importantly, the perspective that underpins this work is one in which native peoples are seen as active participants in changing their own made world; they are shaping their own histories through their *particular* systems of logic, motivation, interests, and meaning. These systems are reflected and played out in everyday events, activities, and social action in which positions are constantly being realigned and renegotiated within specific historic circumstances.

This approach provides a clearer, more colorful and revealing lens through which responses to new materials at the level of individual cultures can be examined. The usefulness of marrying these perspectives with the materials science and ethnohistoric research program devised here is apparent when Illinois reactions to European-derived copper-base metal materials are revealed. The initial decisions the Illinois made to participate in the acquisition, appropriation, and dissemination of foreign copper-base materials were largely made on their own terms. They were clearly not "need-based" in the sense of gaining greater technological "efficiency," or "proficiency" in a technical sense from the objects or the raw materials alone. There is relatively little indication from the material record at the Iliniwek Village that metalworkers were turning European-introduced copper-base metal into utilitarian objects. Yet, these initial choices instigated the reshaping of their material and technological "systems," as well as of their social, economic, and political relations with the changing world around them.

This research has brought to light important contexts in which the Illinois assimilated copper-base metal: 1) reworking it for the purposes of self-expression through enhancement of their appearance (in life and in death?) with body ornamentation; and 2) for use in trade/exchange activities. The former has been demonstrated by the materials analysis aspect of this work, and the other is hypothesized based on the results of the documentary research.

It has been demonstrated clearly from this study, and from the archaeological record, that one of the earliest contexts for material change among the Illinois is the appropriation and local transformation of copper-base trade metals into ornaments and personal adornment items. There seems to be no preference for copper or brass in the manufacture of these ornaments, although the materials were worked differently. Some of these newly reworked items had formal and functional antecedents fashioned in other media, but other forms were new. Their metalworking style, which employs forming techniques known in native North America for millennia, appears less sophisticated than that of their forebears. It certainly departs from the large, impressive, symbolically charged objects of display produced in Mississippian repertoires. Rather, the focus here appears to have been on making small items like clips, beads, and tinkling cones that served as single design elements in the creation of an overall visual, symbolic, and/or aesthetic display. This type of creative social and individual expression surely had the potential to impress. In this sense, the sound, the color, the luster, even the movement of the metals were important qualities that undoubtedly enhanced the overall effect of the design. Thus, European smelted metal was not used in these contexts because it was "superior" in a utilitarian or technological way to other native raw materials or to native copper. It undoubtedly held symbolic meanings and/or aesthetic appeal that transcended its economic utility.

Our knowledge of Illinois involvement in trade and exchange of European items comes primarily from the historic chroniclers rather than from the material record. The level, nature, and potential escalation of their political, social, and economic participation as purported "middlemen" in these activities deserve more attention from scholars, and remain an important hypothesis to be tested. One of the ways archaeologists have begun to think about these questions has been through identifying and tracing the dissemination of traded materials, particular "batches" of copper-base metals, over and across regions. While this requires a significant database of tested materials, studies are underway in this endeavor (Walker et al. 1999).

When considering either of these possibilities, however, it is interesting to note that Illinois copper-base metal consumption patterns indicate that the amount of discarded material calls into serious (and as yet unresolved) question its economic and/or symbolic "worth" during this period. There is no doubt that mortuary data would contribute immeasurably to our understanding of the meaning of these materials to the Illinois as well as their context(s) of use. It would also be quite revealing to know just how much "raw" or transformed material was being traded to other groups. At the same time, however, we can still conclude that procuring, consuming, and exchanging copper-based metals was one important way in which the Illinois set themselves on an early, potentially unique historical trajectory of escalating "entanglement" with Europeans and with other native peoples as foreign influence intensified in the region.

The approaches adopted in this study and the results achieved herein have important implications for the direction of current and further research into questions of native material and culture change in European contact contexts both within and outside of the Western Great Lakes/Upper Mississippi Valley region. Renewed interest in the anthropology of technology has generated provocative new ideas articulating the relations between culture and technology, emphasizing the material, social, and ideological dimensions of making and using things. Many of these have been operationalized here. However, few studies have been generated that demonstrate the applicability of these theories to questions of technological change in specific culture contact contexts.

While my approaches and my findings concerning the timing, contexts, and conditions of early material and technological change in Illinois systems can be related in significant respects to interpretations of the tempo and contexts of material change among the Onondaga (Bradley 1987), the Tionontate Huron (Branstner 1991, 1992), Huron (Anselmi 1994), protohistoric Fort Ancient peoples (Drooker 1996a, b), and Western Lake Erie Basin groups (Stothers 1991, 2000), this research has departed in important ways. Largely following and expanding on Bradley's (1987), Bradley and S. Terry Childs' (1991), and Rita Wright's (1989, 1993) lead, this research makes explicit the meanings and contexts of Illinois acceptance and integration of one particular type of introduced material within a particular, albeit already changing cultural system. This is accomplished using a view of technology in which all technological activity is viewed as potentially innovative, creative, and imbued with multiple levels of social and ideological significance which can be revealed through detailed, fine-grained, finely textured analysis. From this perspective, this research has challenged "normative" explanations of material change which assume that the early appearance of European goods in native material culture assemblages had "minimal" or "negligible" effect on native technological systems because "new" items were seen to have been "functionally equivalent" to native forms or simply "reshaped in traditional ways" and "integrated into existing systems." As has been demonstrated in this work, such explanations deny the dynamic nature of technological systems and fail to bring into focus what has been termed the microenvironments of change, that is, the local, everyday responses that illuminate potentially important reactions to European material influence.

Theoretically and methodologically, this research has contributed to and has direct relevance to future research endeavors related to the structure of native metalworking systems, and continuity and change within them over time and space. Jonathan Leader (1988) and Childs (1994) have called for further technical and archaeometric analyses of prehistoric Native American metals assemblages to illuminate such processes through prehistory. This work has demonstrated that while primary metalworking techniques persisted through the early contact period, secondary techniques diverged significantly from previous systems, resulting in a new technological "style" in design, fabrication, use, and meaning. From a technical and ideological perspective, examining the transition from working native, then smelted European copper and brass materials is crucial to our understanding of variation, standardization, and innovation in materials selection, use, and working methods both within technological systems and among them. The materials-based, archaeometric investigation conducted here using a combined suite of techniques allows archaeologists to comprehend aspects of metalworking practices that are inaccessible by stylistic and formal evaluation alone. This is an integral aspect of technological inquiry and contributes significantly to these understandings. In this way, the dynamic nature of native metalworking systems over the prehistoric/historic transition and thereby over the *longue durée* of history is truly revealed. Continuing large-sample, materials-oriented research into copper-base metal composition, distribution, consumption, and working techniques among protohistoric and early historic peoples is one major way in which important links between protohistoric peoples of the interior and peoples of the Northeast where processes of transformation were already well in place can be accessed.

This research has rendered a fine grained, richly textured treatment of the ways in which one group, the seventeenth-century Illinois, responded technologically to the availability of a particular newly available material—smelted copper-base metal—within a culture contact setting. The perspectives on technological style and on technology as an inclusive system, combined with analytical approaches and archaeometallurgical methods that illuminate the material, contextual, and historical dimensions of making and using things adopted here have opened important new windows on the dynamic relationships among technology and culture, the meanings of material culture change, and Europeans and Native Americans in this poorly understood period of native cultural upheaval. It is hoped that the results obtained here stimulate further comparative investigations into the timing and trajectory of material, technological, and social change among native groups as European influence intensifies during the contact period. In this way, the potentially unique ways in which individual native culture groups responded to the conjunctures of long and short term history will be illuminated.

Notes

CHAPTER 2

1. Noel Perrin (1979) chronicles the ready acceptance, then ultimate rejection of guns by the Japanese. First introduced by the Portuguese in 1543, within ten years of their initial adoption, expert Japanese swordsmiths had mastered the technology to build and use them. Despite their superiority as weapons, firearms were abandoned by the mid-seventeenth century, primarily because they did not fit into systems of honor, dignity, and warfare as symbolized and practiced by skilled samurai swordsmen of the warrior class.

CHAPTER 3

1. "Context" is defined here after Ian Hodder (1986:120–121) who states that context relates to the "connecting or interweaving of things in a particular situation or group of situations" so as to be meaningful in their functional interrelationships and/or in the structured content of ideas and symbols. As Hodder (1986:139) points out, artifacts exist in many dimensions, or relevant environments, at once. Another useful discussion concerning the importance of "context" in interpreting postcontact material culture change can be found in Rogers (1990:15–16).

2. Many of these studies are discussed in greater detail in chapter 4. A comprehensive history of archaeometallurgical inquiry into native North American metalworking practices may be found in Susan Martin (1999).

3. Early studies by Wilson and Sayre (1935) and Drier (1961), are cited in chapter 4. They will not be reviewed here.

4. It should be noted that two of the three studies cited remain unpublished.

5. Friability caused by deeply penetrating corrosion (embrittlement) was a major concern. As Scott (1991:61) has pointed out, friable materials can break easily, even crumble along lines of weakness, causing damage to the artifact. They can also cause damage to the blade of the cutting saw used to remove the sample.

6. Only two samples, both double-lobed tubing, were too brittle and fragmentary for INAA. Both were eliminated from the INAA data set.

7. The term "document" is defined here as "any material that includes a written message or depiction," including "handwritten or printed written texts, scrawled notes, drawn maps or pictures, paintings, photographs, and anything else designed to communicate a message visually" (Barber and Berdan 1998:29).

CHAPTER 4

1. Halsey (1996) and Martin (1999) provide excellent syntheses of modern and historic efforts.

2. Day (1998) and Pollard and Heron (1996) present detailed histories of medieval brass production.

3. The term *bending* is used here instead of *rolling* so as not to misuse the term *rolling,* which in metallurgical terms, refers to the metal-forming operation in which the thickness (cross-sectional area) of sheet stock is reduced by running it between rotating rollers (Callister 2000:838–839; Chandler 1998:250). The term *rolling* is, however, commonly used by protohistorians when describing metal shaping techniques in which metal is bent in a curved manner, as around a mandrel. In this sense, *rolling* is a form of bending.

4. Laurier Turgeon's (1997:9) translation of the first part of this relation is "men and women alike wore 'ear pendants in the manner of the Orientals, *particularly strips of chiseled copper*' " (emphasis mine). This translation provides clearer indication that native peoples were found wearing copper ornaments that they had fashioned themselves of native copper.

5. Eugenia Herbert (1984:277–282) has made similar observations concerning the symbolic associations of the red color of copper with life, blood, and power among African peoples.

6. These techniques are discussed in greater detail in chapters 6 and 8.

7. See also Mary Helms (1993) for the special values and qualities imbued in objects and materials acquired from a distance.

CHAPTER 5

1. Chapter title quotation from Jesuit Father Claude Dablon's entry in the Jesuit *Relations* of 1669–1671 (Thwaites 1896–1901:54:187–191).

2. It is important to emphasize that the term "middlemen" as it is commonly

understood (especially by Europeans of the time) may have limited utility for explaining adequately the diverse motives, actions, and goals of those native peoples who engaged in and attempted to control the dissemination of foreign items and materials into the hinterlands. White (1991:105–106) questions the very notion of an exclusive "middleman" status, pointing to the political, economic, and social complexities of the conditions, contexts, and relations of exchange among the numerous groups involved in it.

3. Numerous authors have emphasized that while the Illinois tribes are often referred to as a "confederacy," they were not as tightly organized politically as the Iroquois or the Creek (Bauxar 1978:594; Callender 1978:673). Margaret Brown (1979:233) suggests that the designation may have applied only during the latter historic period, when the Illinois were depopulated and were negotiating treaties with other Algonquian groups.

4. French explorer Etienne Brulé may have actually seen Lakes Michigan and Superior as early as 1622, but reports of the purported voyage are third hand and cannot be confirmed (Brose 1971:52).

5. Among others, Good (1972) and Balesi (1992) discuss the notion that the journal attributed to Marquette was actually written after the 1673 voyage by Father Claude Dablon based on Jolliet's recollections. Temple (1966) also alludes to its compilation after the voyage. Father Lucien Campeau (1991) has made an exhaustive, critical study of the problem of authenticity, and has argued conclusively that the narration is indeed authored by Father Marquette himself.

6. *8* is a symbol common to all French missionary writing systems on the Illinois language, with some variations of use according to individual recorders. It represents phonemic vowels *o, oo,* and *w* (Costa 2002:3–4, 2003: 12–14).

7. The Ottawa are thought to have succeeded the Huron as major "middlemen" in the Western Great Lakes trade. As already pointed out in note 2 above, White (1991:106) questions the exclusivity of this position. He also postulates that the Ottawa's hegemonic activities actually expanded rather than limited trade to the interior.

8. It is important to remember that despite increasing French commercial, military, and missionary presence in the Illinois Country, the first settlers did not arrive in the region until after 1718 (Walthall and Emerson 1991:5–6).

9. The French were interested in exploiting the Illinois Country, but were reluctant to arm the Illinois heavily.

10. Speculation as to the exact location of the site lasted nearly a century. Over the years, researchers have variously postulated that the village was located on the Iowa, Des Moines, or even the Missouri River (Temple 1966:15–19).

11. Rochelle Lurie (1998) reports that copper-base trade metal was found at the New Lenox site, Will County, Illinois, in contexts radiocarbon dated to early in the third decade of the seventeenth century.

12. In her seminal ethnohistoric study of Illinois depopulation, Emily Blasingham (1956:373) points to raiding, disease, introduction of Christianity and monogamy, excessive liquor consumption, and group fissioning as factors contributing to depopulation. How early on in the contact process these factors, especially disease, came into play is unknown.

CHAPTER 6

1. To avoid redundancy in the use of the term "copper-base metal" throughout the next two chapters, the term "metal" will be taken to mean "copper-base" metal.

2. Historian Jan Kupp (1968:100) points out that in 1626, Dutch merchants offered "French Indians" high quality copper kettles that were far too heavy to be transported by canoe or over narrow trails.

3. These are *not* the same spiral forms to which Bradley and Childs (1991) and Wray et al. (1987:59–60) refer.

4. These forms are used by many native peoples in the present, but care must be taken not to assume that the history of their use has been continuous or that their production technology, use contexts, or symbolic meanings have remained the same.

5. Iron tinkling cones also appear during the contact period, but much less frequently. Three were recovered from the 1947 work at the Zimmerman site (Brown 1975:32).

6. As it pertains here, a "seam" is defined in the following manner: "On the surface of metal, an unwelded fold or lap that appears as a crack, usually resulting from a defect obtained in casting or in working" (American Society for Metals Committee on Definitions of Metallurgical Terms 1977:44).

CHAPTER 7

1. In addition to having adequate precisional capabilities, it also measures lead, a critical diagnostic element in provenance studies of smelted materials (Fleming and Swann 2000:712). INAA does not measure lead.

2. It should be remembered that PIXE did not detect any absolute values for zinc in specimens placed in the European-introduced copper group.

3. Investigators have demonstrated that changes in the microstructure of cartridge brass (70/30) are first detectable when the metal is reduced more than 10 percent (ASM Handbook Committee 1972:281–285; Samans 1963:59–63)

4. In their INAA study of the native and European copper-base metals from Ontario, Hancock, Farquhar, Pavlish, and Finlayson (1995:258–261) include a category called "brassy copper" in which they place zinc-rich copper artifacts having ≤ 8 percent zinc. They place in this category artifacts with as little as 1.1 percent zinc. There are no specimens in this study that fit in this category.

5. Melting was known among the Illinois for other materials, however. Brown (1975:33) reports that fragments of objects created by grinding glass beads into a powder, making a paste, and heating it on a piece of brass sheet were recovered from the Zimmerman site. Melted glass pendants were also found at several eighteenth-century sites attributable to the Illinois (Brown 1972).

CHAPTER 8

1. See notes 2 and 7, chapter 5.

2. One important point of interethnic recognition and cultural convergence between native peoples and the French was their common respect and admiration for rituals and ceremony. Highly structured, ritualized greetings and behaviors accompanied by mutual recognition of rank and prestige through costume, gesture, and presentation of gifts were important components of interactions and material exchange (Jaenen 1974:278–280; Seed 1995).

3. In Powhatan society, a *weroance* is a district and/or petty chief or literally, a commander (Roundtree 1989:16). Powhatan himself was the paramount chief.

References Cited

Abel, Timothy J., Adrian L. Burke, and David M. Stothers

2000 The Protohistoric Transition in the Southwestern Lake Erie Region. Paper presented at the 33rd Annual Meeting of the Canadian Archaeological Association, Ottawa.

American Society for Metals (ASM) Committee on Definitions of Metallurgical Terms

1977 *Definitions of Metallurgical Terms.* American Society for Metals, Materials Park, Ohio.

American Society for Metals (ASM) Handbook Committee

1972 *Atlas of Microstructures of Industrial Alloys.* Metals Handbook, Vol. 7, edited by Taylor Lyman, pp. 282–285. 8th edition. American Society for Metals, Materials Park, Ohio.

Anselmi, Lisa M.

1994 *Recycle and Rework: Huron Manipulation of European-Introduced Metal Trade Goods at the Auger Site (BdGw-3), Simcoe County, Ontario.* M.Sc. Thesis, Department of Anthropology, University of Toronto.

Anselmi, Lisa, Martha Latta, and R. G. V. Hancock

1997 Instrumental Neutron Activation Analysis of Copper and Brass from the Auger Site (BDGW-3), Simcoe County, Ontario. *Northeast Anthropology* 53:47–59.

Armstrong, Douglas V.

2001 Attaining the Full Potential of Historical Archaeology. *Historical Archaeology* 35(2):1–8.

Balesi, Charles

1992 *The Time of the French in the Heart of North America, 1673–1818.* Alliance Française, Chicago.

Barber, Russell J., and Frances F. Berdan
1998 *The Emperor's Mirror: Understanding Culture through Primary Sources.* University of Arizona Press, Tucson.

Basalla, George
1988 *The Evolution of Technology.* Cambridge University Press, Cambridge.

Bastian, Tyler
1961 Trace Element and Metallographic Studies of Prehistoric Copper Artifacts in North America: A Review. In *Lake Superior Copper and the Indians: Miscellaneous Studies of Great Lakes Prehistory,* edited by James B. Griffin, pp. 151–175. Anthropological Papers 17–18. Museum of Anthropology, University of Michigan, Ann Arbor.

Bauxar, J. Joseph
1978 History of the Illinois Area. In *Northeast,* edited by Bruce G. Trigger, pp. 594–601. Handbook of North American Indians, Vol.15, William G. Sturtevant, general editor. Smithsonian Institution, Washington, D.C.

Bayley, Justine
1998 The Production of Brass in Antiquity with Particular Reference to Roman Britain. In *2000 Years of Zinc and Brass,* edited by P. T. Craddock, pp. 7–26. 1st revised edition. Occasional Papers 50. British Museum, London.

Beaulieu, Alain, and Roland Viau
2001 *The Great Peace: Chronicle of a Diplomatic Saga.* Éditions Libre Expression, Montreal.

Bender, Barbara
1989 The Roots of Inequality. In *Domination and Resistance,* edited by Daniel Miller, Michael Rowlands, and Christopher Tilley, pp. 83–95. Unwin Hyman, London.

Binford, Lewis
1962 Archaeology as Anthropology. *American Antiquity* 28:217–225.
1965 Archaeological Systematics and the Study of Culture Process. *American Antiquity* 31:203–210.

Birk, Douglas, and Elden Johnson
1992 The Mdewakanton Dakota and Initial French Contact. In *Calumet and Fleur-de-Lys: Archaeology of Indian and French Contact in the Midcontinent,* edited by John A. Walthall and Thomas E. Emerson, pp. 203–240. Smithsonian Institution Press, Washington, D.C.

Bisson, Michael S.
2000 Precolonial Copper Metallurgy: Sociopolitical Context. In *Ancient African Metallurgy: The Socio-Cultural Context,* edited by Joseph O. Vogel, pp. 83–145. Altamira Press, Walnut Creek, California.

Blair, Emma Helen
1996 *The Indian Tribes of the Upper Mississippi Valley and Region of the Great*

Lakes, Vols. 1 and 2. Reprinted. University of Nebraska Press, Lincoln. Volume 1 originally published 1911, Arthur H. Clark Co., Cleveland. Volume 2 originally published 1912, Arthur H. Clark Co., Cleveland.

Blasingham, Emily

1956 The Depopulation of the Illinois Indians. *Ethnohistory* 3(3–4): 224–293, 361–412.

Bleed, Peter, Dennis L. Toom, Craig M. Johnson, Carlyle S. Smith, and Ricky Roberts

1995 Metal Artifacts and Other European Trade Materials. In *Archeology of the Medicine Crow Site Complex (39BF2), Buffalo County, South Dakota,* edited by Stanley A. Ahler and Dennis L. Toom, pp. 325–342. Reports of Investigations 51. Illinois State Museum, Springfield.

Bradley, James W.

1987 *Evolution of the Onondaga Iroquois: Accommodating Change, 1500–1655.* Syracuse University Press, Syracuse.

Bradley, James W., and S. Terry Childs

1991 Basque Earrings and Panther's Tails: The Form of Cross-Cultural Contact in Sixteenth-Century Iroquoia. In *Metals in Society: Theory beyond Analysis,* edited by Robert M. Ehrenreich, pp. 7–17. Research Papers in Science and Archaeology, Vol. 8, Part 2. MASCA, University Museum of Archaeology and Anthropology, University of Pennsylvania, Philadelphia.

Brady, George S., Henry R. Clauser, and John A. Vaccari

1997 *Materials Handbook.* 14th edition. McGraw Hill, New York.

Brain, Jeffrey

1979 *Tunica Treasure.* The Peabody Museum of Archaeology and Ethnology, Harvard University and the Peabody Museum of Salem, Cambridge.

Branstner, Susan

1991 Decision-Making Processes in a Culture Contact Context: The Case of the Tionontate Huron of the Upper Great Lakes. Ph.D. dissertation, Michigan State University. University Microfilms, Ann Arbor.

1992 Tionontate Huron Occupation at the Marquette Mission. In *Calumet and Fleur-de-Lys: Archaeology of Indian and French Contact in the Midcontinent,* edited by John A. Walthall and Thomas E. Emerson, pp. 177–201. Smithsonian Institution Press, Washington, D.C.

Braudel, Fernand

1972 *The Mediterranean and the Mediterranean World in the Age of Phillip II,* Vol. 1. Translated by Siam Reynolds. Originally published 1949. Harper Colophon, New York.

Braund, Kathryn. E. Holland

1993 *Deerskins and Duffels: Creek Indian Trade with Anglo-America, 1685–1815.* University of Nebraska Press, Lincoln.

Bray, Robert
1978 European Trade Goods from the Utz Site and the Search for Fort Orleans. *Missouri Archaeologist* 13(1–75).

Brissaud, I., G. LaGarde, A. Sabir, and A. Houdayer
1987 PIXE Macro and Microprobe Techniques in Archaeometry. *Journal of Radioanalytical and Nuclear Chemistry* 116(1):99–116.

Brose, David S.
1971 The Direct Historic Approach to Michigan Archaeology. *Ethnohistory* 18(1): 51–61.

Brown, Ian
1979 Historic Artifacts and Sociocultural Change: Some Warnings from the Lower Mississippi Valley. In *The Conference on Historic Site Archaeology Papers 1978,* 13:109–121. Institute of Archeology and Anthropology, University of South Carolina, Columbia.

Brown, James A.
1961 *The Zimmerman Site: A Report on Excavations at the Grand Village of Kaskaskia, LaSalle County, Illinois.* Reports of Investigations, No. 9. Illinois State Museum, Springfield.
1985 The Mississippian Period. In *Ancient Art of the American Woodland Indians,* edited by D. Brose, J. Brown, and D. Penney, pp. 93–140. Harry N. Abrams, New York.
1989 On Style Divisions of the Southeastern Ceremonial Complex: A Revisionist Perspective. In *The Southeastern Ceremonial Complex: Artifacts and Analysis,* edited by Patricia Galloway, pp. 183–204. University of Nebraska Press, Lincoln.
1990a Ethnohistoric Connections. In *At the Edge of Prehistory: Huber Phase Archaeology in the Chicago Area,* edited by James A. Brown and Patricia J. O'Brien, pp. 155–160. Center for American Archeology, Kampsville, Illinois.
1990b Ground Stone, Metallic, and Glass Artifacts. In *At the Edge of Prehistory: Huber Phase Archaeology in the Chicago Area,* edited by James A. Brown and Patricia J. O'Brien, pp. 236–240. Center for American Archeology, Kampsville, Illinois.

Brown, James A., Richard A. Kerber, and Howard D. Winters
1990 Trade and the Evolution of Exchange Relations at the Beginning of the Mississippian Period. In *The Mississippian Emergence,* edited by B. D. Smith, pp. 251–280. Smithsonian Institution Press, Washington, D.C.

Brown, James A., and Robert Sasso
2001 Prelude to History on the Eastern Prairies. In *Societies in Eclipse: Archaeology of the Eastern Woodlands Indians, A.D. 1400–1700,* edited by David S. Brose, C. Wesley Cowan, and Robert C. Mainfort Jr., pp. 205–228. Smithsonian Institution Press, Washington, D.C.

Brown, Margaret K.

1972 Native Made Glass Pendants from East of the Mississippi. *American Antiquity* 37(3):432–439.

1975 *The Zimmerman Site: Further Investigations at the Grand Village of Kaskaskia.* Reports of Investigations No. 32. Illinois State Museum, Springfield.

1979 Cultural Transformations among the Illinois: An Application of a Systems Model. Anthropological Series 1(3). Publications of the Museum, Michigan State University, East Lansing.

Callender, Charles

1978 Illinois. In *Northeast,* edited by Bruce G. Trigger, pp. 673–680. Handbook of North American Indians, Vol. 15, William C. Sturtevant, general editor, Smithsonian Institution, Washington, D.C.

Callister, William D.

2000 *Materials Science and Engineering: An Introduction.* 5th edition. John Wiley and Sons, New York.

Campeau, Lucien, S.J., S.R.C.

1991 Regard critique sur la *Narration* du P. Jacques Marquette. *Les Cahiers des Dix* 46:21–60.

Cannon, Aubrey

1991 Gender, Status, and the Focus of Material Display. In *The Archaeology of Gender,* edited by Dale Walde and Noreen D. Willows, pp. 144–149. Proceedings of the 22nd Annual Chacmool Conference. The Archaeological Association of the University of Calgary, Calgary.

Chandler, Harry

1998 *Metallurgy for the Nonmetallurgist.* American Society for Metals International, Materials Park, Ohio.

Chapman, Jefferson

1985 *Tellico Archeology: 12,000 Years of Native American History.* University of Tennessee Press, Knoxville.

Childs, S. Terry

1991a Iron as Utility or Expression: Reforging Function in Africa. In *Metals in Society: Theory beyond Analysis,* edited by Robert M. Ehrenreich, pp. 7–17. Research Papers in Science and Archaeology, Vol. 8, Part 2. MASCA, University Museum of Archaeology and Anthropology, University of Pennsylvania, Philadelphia.

1991b Style, Technology, and Iron Smelting Furnaces in Bantu-Speaking Africa. *Journal of Anthropological Archaeology* 10:332–359.

1994 Native Copper Technology and Society in Eastern North America. In *Archaeometry of Pre-Columbian Sites and Artifacts,* edited by David A. Scott and Pieter Meyers, pp. 229–253. Getty Conservation Institute, Los Angeles.

Ciliberto, Enrico

2000 Analytical Methods in Art and Archaeology. In *Modern Analytical Methods in Art and Archaeology,* edited by Enrico Ciliberto and Giuseppe Spoto, pp. 1–10. John Wiley and Sons, New York.

Cleland, Charles E.

1971 Metallic Artifacts. In *The Lasanen Site: An Historic Burial Locality in Mackinac County, Michigan,* edited by Charles E. Cleland, pp. 19–34. Anthropological Series 1(1). Publications of the Museum, Michigan State University, East Lansing.

1992 *Rites of Conquest: The History and Culture of Michigan's Native Americans.* University of Michigan Press, Ann Arbor.

2001 Historical Archaeology Adrift? *Historical Archaeology* 35(2):1–8.

Cobb, Charles R.

1991 Social Reproduction and the *Longue Durée* in the Prehistory of the Midcontinental United States. In *Processual and Postprocessual Archaeologies: Multiple Ways of Knowing the Past,* edited by Robert W. Preucel, pp. 168–182. Occasional Paper No.10. Center for Archaeological Investigations, Southern Illinois University, Carbondale.

Conkey, Margaret

1989 The Structural Analysis of Paleolithic Art. In *Archaeological Thought in America,* edited by C. C. Lamberg-Karlovsky, pp. 135–154. Cambridge University Press, Cambridge.

Cordell, Linda, and Vincent J. Yannie

1991 Ethnicity, Ethnogenesis, and the Individual: A Processual Approach toward Dialogue. In *Processual and Postprocessual Archaeologies: Multiple Ways of Knowing the Past,* edited by Robert W. Preucel, pp. 96–107. Occasional Paper No. 10. Center for Archaeological Investigations, Southern Illinois University, Carbondale.

Costa, David

2002 An Overview of the Illinois Language. In *Kaskaskia Illinois-to-French Dictionary,* edited by Carl Masthay, pp. 2–4. Published by the author, St. Louis.

2003 *The Miami-Illinois Language.* University of Nebraska Press, Lincoln.

Craddock, Paul T.

1995 *Early Metal Mining and Production.* Smithsonian Institution Press, Washington, D.C.

Cushing, Daniel

1967 Principles of Corrosion Applicable to Ancient Metals and Methods of Identifying Corrosion Products. In *Application of Science in Examination of Works of Art,* edited by W. J. Young, pp. 53–65. Museum of Fine Arts, Boston.

Cushing, Frank

1894 Primitive Copper Working: An Experimental Study. *American Anthropologist* (Old Series) 7:93–117.

Cusick, James G.

1998 The Historiography of Acculturation: An Evaluation of Concepts and Their Application in Archaeology. In *Studies in Culture Contact: Interaction, Culture Change, and Archaeology,* edited by James G. Cusick, pp. 126–145. Occasional Paper No. 25. Center for Archaeological Investigations, Southern Illinois University, Carbondale.

Day, Joan

1998 Brass and Zinc in Europe from the Middle Ages until the Nineteenth Century. In *2000 Years of Zinc and Brass,* edited by P. T. Craddock, pp.133–158. 1st revised edition. Occasional Papers No. 50. British Museum, London.

Deagan, Kathleen A.

1988 Neither History Nor Prehistory: The Questions That Count in Historical Archaeology. *Historical Archaeology* 22(1):7–12.

1998 Transculturation and Spanish American Ethnogenesis: The Archaeological Legacy of the Quincentenary. In *Studies in Culture Contact: Interaction, Culture Change, and Archaeology,* edited by James G. Cusick, pp. 23–43. Occasional Paper 25. Center for Archaeological Investigations, Southern Illinois University, Carbondale.

Dietler, Michael

1998 Consumption, Agency, and Cultural Entanglement: Theoretical Implications of a Mediterranean Colonial Encounter. In *Studies in Culture Contact: Interaction, Culture Change, and Archaeology,* edited by James G. Cusick, pp. 288–315. Occasional Paper No. 25. Center for Archaeological Investigations, Southern Illinois University, Carbondale.

Dietler, Michael, and Ingrid Herbich

1998 *Habitus,* Techniques, Style: An Integrated Approach to the Social Understanding of Material Culture and Boundaries. In *The Archaeology of Social Boundaries,* edited by Miriam Stark, pp. 232–263. Smithsonian Institution Press, Washington, D.C.

Dobres, Marcia-Anne

2001 Meaning in the Making: Agency and the Social Embodiment of Technology and Art. In *Anthropological Perspectives on Technology,* edited by Michael Brian Schiffer, pp. 47–76. University of New Mexico Press, Albuquerque.

Dobres, Marcia-Anne, and Christopher R. Hoffman

1994 Social Agency and the Dynamics of Prehistoric Technology. *Journal of Archaeological Method and Theory* 1(3):211–258.

1999 Introduction: A Context for the Present and Future of Technology Studies.

In *The Social Dynamics of Technology: Practice, Politics, and World Views,* edited by Marcia-Anne Dobres and Christopher R. Hoffman, pp. 1–19. Smithsonian Institution Press, Washington, D.C.

van Dongen, Alexandra

1996 The Inexhaustible Kettle: The Metamorphosis of a European Utensil in the World of the North American Indians. In *One Man's Trash Is Another Man's Treasure,* edited by Alexandra van Dongen, pp. 115–171. Museum Boymans-van Beuningen, Rotterdam.

Donnelly, Joseph P., S.J.

1968 *Jacques Marquette, S.J.* Loyola University Press, Chicago.

Dran, Jean-Claude, Thomas Calligaro, and Joseph Salomon

2000 Particle-Induced X-ray Emission. In *Modern Analytical Methods in Art and Archaeology,* edited by Enrico Ciliberto and Giuseppe Spoto, pp. 135–166. John Wiley and Sons, New York.

Drier, Roy Ward

1961 Archaeology and Some Metallurgical Techniques. In *Lake Superior Copper and the Indians: Miscellaneous Studies of Great Lakes Prehistory,* edited by James B. Griffin, pp.134–147. Anthropological Papers No. 17. Museum of Anthropology, University of Michigan, Ann Arbor.

Drooker, Penelope B.

1993 Chemical Analysis of Metal Samples from the Madisonville Site. Report to the Cincinnati Museum of Natural History, February.

1996a *The View from Madisonville: Continuity and Change in Late Prehistoric-Protohistoric Western Fort Ancient Interaction Patterns.* Ph.D. dissertation, University at Albany, State University of New York. University Microfilms, Ann Arbor.

1996b Madisonville Metal and Glass Artifacts: Implications for Western Fort Ancient Chronology and Interaction Patterns. *Midcontinental Journal of Archaeology* 21(2):145–190.

Drooker, Penelope B., and C. Wesley Cowan

2001 Transformation of the Fort Ancient Cultures of the Central Ohio Valley. In *Societies in Eclipse: Archaeology of the Eastern Woodlands Indians, A.D. 1400–1700,* edited by David S. Brose, C. Wesley Cowan, and Robert C. Mainfort Jr., pp. 83–106. Smithsonian Institution Press, Washington, D.C.

Dubin, Lois Sherr

1999 *North American Indian Jewelry and Adornment: From Prehistory to the Present.* Harry N. Abrams, New York.

Dunbar, Helene R., and Katharine C. Ruhl

1974 Copper Artifacts from the Engelbert Site. *The Bulletin* 61:1–10. New York State Archeological Association.

Dymond, D. P.

1974 *Archaeology and History: A Plea for Reconciliation.* Thames and Hudson, London.

Earle, Timothy

1991 Toward a Behavioral Archaeology. In *Processual and Postprocessual Archaeologies: Multiple Ways of Knowing the Past,* edited by Robert W. Preucel, pp. 83–95. Occasional Paper No. 10. Center for Archaeological Investigations, Southern Illinois University, Carbondale.

Ehrenreich, Robert M. (editor)

1991 *Metals in Society: Theory Beyond Analysis.* Research Papers in Science and Archaeology, Vol. 8, Part 2. MASCA, University Museum of Archaeology and Anthropology, University of Pennsylvania, Philadelphia.

Ehrhardt, Kathleen L.

2002 European Materials in Native American Contexts: Rethinking Technological Change. Unpublished Ph.D. dissertation, Department of Anthropology, New York University, New York.

2004 Linking History and Prehistory in the Midcontinent: Archaeological Investigations at Marquette and Jolliet's "Peouarea." Papers in honor of Dr. Howard D. Winters, edited by Anne-Marie Cantwell and Lawrence Conrad. Joint publication of the Illinois State Museum and the Center for American Archaeology, Springfield.

Ehrhardt, Kathleen L., and Lawrence A. Conrad

1994 Recent Investigations at the Iliniwek Village Historic Site, Clark County, Missouri. Paper presented at the Joint 51st Southeastern Archaeological Conference and 39th Midwest Archaeological Conference, Lexington, Kentucky.

Ehrhardt, Kathleen L., and Larry D. Grantham

2001 Illinois/Chiwere Siouan Interaction in the Late Protohistoric Midcontinent. Paper presented at the 47th Midwest Conference, La Crosse.

Ehrhardt, Kathleen L., Samuel K. Nash, and Charles P. Swann

2000 Metal-Forming Practices among the Seventeenth Century Illinois. *Materials Characterization* 45:275–288.

Emerson, Thomas E., and James A. Brown

1992 The Late Prehistory and Protohistory of Illinois. In *Calumet and Fleur-de-Lys: Archaeology of Indian and French Contact in the Midcontinent,* edited by John A. Walthall and Thomas E. Emerson, pp. 77–128. Smithsonian Institution Press, Washington, D.C.

Emerson, Thomas E., and Floyd Mansberger

1991 The Search for French Peoria. In *French Colonial Archaeology: The Illinois Country and the Western Great Lakes,* edited by John A. Walthall, pp. 149–164. University of Illinois Press, Urbana.

Fitting, James

1976 Patterns of Acculturation at the Straits of Mackinac. In *Cultural Change and Continuity: Essays in Honor of James Bennett Griffin,* edited by Charles Cleland, pp. 321–334. Academic Press, New York.

1980 Te Oskonchiae: The Village of the Tionontate Huron at Michilimackinac, 1671–1701. Ms. on file, Bureau of History, Michigan Department of State, Lansing.

Fitzgerald, William R.

1990 Chronology to Culture Process: Lower Great Lakes Archaeology, 1500–1650. Unpublished Ph.D. dissertation, Department of Anthropology, McGill University, Montreal.

1995 A Late Sixteenth-Century European Trade Assemblage from Northeastern North America. In *Trade and Discovery: The Scientific Study of Artefacts from Post-Medieval Europe and Beyond,* edited by Duncan R. Hook and David R. M. Gaimster, pp. 29–44. Occasional Paper No. 109. British Museum, London.

Fitzgerald, William R., and Peter G. Ramsden

1988 Copper Based Metal Testing as an Aid to Understanding Early European-Amerindian Interaction: Scratching the Surface. *Canadian Journal of Archaeology* 12:153–161.

Fitzgerald, William R., Laurier Turgeon, R. H. Whitehead, and James Bradley

1993 Late Sixteenth-Century Basque Banded Copper Kettles. *Historical Archaeology* 27(1):44–57.

Fitzhugh, William

1985 Introduction. In *Cultures in Contact: The European Impact on Native Cultural Institutions in Eastern North America, A.D. 1000–1800,* edited by William W. Fitzhugh, pp.1–15. Smithsonian Institution Press, Washington, D.C.

Fleming, S. J., and C. P. Swann

1993 Recent Applications of PIXE Spectrometry in Archaeology Part 1: Observations on the Early Development of Copper Metallurgy in the Old World. *Nuclear Instruments and Methods in Physics Research* B75:440–444.

2000 Distinguishing between Native and Smelted Coppers Using PIXE Spectrometry: A Case History from Early Colonial America. *Nuclear Instruments and Methods in Physics Research* B161–163:709–713.

Fleming, S. J., C. P. Swann, P. E. McGovern, and L. Horne

1990 Characterization of Ancient Materials Using PIXE Spectrometry. *Nuclear Instrumentation and Methods in Physics Research* B49:203–299.

Fowler, Loretta

1987 *Shared Symbols and Contested Meanings: Gros Ventre Culture and History, 1778–1984.* Cornell University Press, Ithaca.

Fox, William A.

1991 The Serpent's Copper Scales. *Wanikan* 91–03:3–11.

1992 Dragon Sideplates from York Factory: A New Twist on an Old Tail. *Manitoba Archaeological Journal* 2(2):21–35.

Fox, William A., R. G. V. Hancock, and L. A. Pavlish

1995 Where East Met West: The New Copper Culture. *Wisconsin Archeologist* 76(3–4):269–293.

Franke, Judith A.

1995 *French Peoria and the Illinois Country, 1673–1846.* Illinois State Museum Popular Science Series XII. Illinois State Museum Society, Springfield.

Franklin, Ursula M., E. Badone, R. Gotthardt, and B. Yorga

1981 *An Examination of Prehistoric Copper Technology and Copper Sources in Western Arctic and Subarctic North America.* National Museum of Man Mercury Series Paper No. 101. National Museums of Canada, Ottawa.

Gallay, Alan

2002 *The Indian Slave Trade: The Rise of the English Empire in the American South, 1670– 1717.* Yale University Press, New Haven.

Galloway, Patricia

1991 The Archaeology of Ethnohistoric Narrative. In *The Spanish Borderlands in Pan-American Perspective,* edited by David Hurst Thomas, pp. 453–469. Columbian Consequences, Vol. 3. Smithsonian Institution Press, Washington, D.C.

Gersch, H. K., J. D. Robertson, A. G. Henderson, D. Pollack, and C. A. Munson

1998 PIXE Analysis of Prehistoric and Protohistoric Caborn-Welborn Phase Copper Artifacts from the Lower Ohio River Valley. *Journal of Radioanalytical and Nuclear Chemistry* 234(1–2):85–90.

Gibbon, Guy

1995 Oneota at the Periphery: Trade, Political Power, and Ethnicity in Northern Minnesota and on the Northeastern Plains in the Late Prehistoric Period. In *Oneota Archaeology, Past, Present, and Future,* edited by William Green, pp. 175–199. Report 20. Office of the State Archaeologist, University of Iowa, Iowa City.

1998 Old Copper in Minnesota: A Review. *Plains Anthropologist* 43(163):27–50.

Glascock, Michael D.

1992 Characterization of Archaeological Ceramics at MURR by Neutron Activation Analysis and Multivariate Statistics. In *Chemical Characterization of Ceramic Pastes in Archaeology,* edited by Hector Neff, pp. 11–26. Prehistory Press, Madison.

Glascock, Michael D., Hector Neff, James W. Cogswell, and R. Sergio Herrera (compilers)

1996 *Neutron Activation Analysis for Archaeological Applications: The Missouri Uni-*

versity Research Reactor Archaeometry Laboratory. Missouri University Research Reactor, Columbia.

Glassie, Henry

1988 Meaningful Things and Appropriate Myths: The Artifact's Place in American Studies. In *Material Life in America, 1600–1860,* edited by Robert Blair St. George, pp. 63–92. Northeastern University Press, Boston.

Goad, Sharon

1979 Middle Woodland Exchange in the Prehistoric Southeastern United States. In *Hopewell Archaeology: the Chillicothe Conference,* edited by David S. Brose and N'omi Greber, pp. 239–246. Kent State University Press, Kent, Ohio.

1980 Chemical Analysis of Native Copper Artifacts from the Southeastern United States. *Current Anthropology* 21(2):270–271.

Goad, Sharon, and John Noakes

1978 Prehistoric Copper Artifacts in the Eastern United States. *Archaeological Chemistry,* Vol. II, edited by Giles Carter, pp. 335–46. Advancements in Chemistry Series, American Chemical Society, Washington, D.C.

Good, Mary Elizabeth

1972 Guebert Site: An 18th Century, Historic Kaskaskia Indian Village in Randolph County, Illinois. Memoir No. 2. Central States Archaeological Societies, St. Louis.

Grantham, Larry D.

1993 The Illini Village of the Marquette and Jolliet Voyage of 1673. *Missouri Archaeologist* 54:1–20.

1994 Excavations at the Haas Site: One of the Peoria Villages of the Jolliet and Marquette Expedition of 1673. Paper presented at the 1994 Annual Workshop of the Illinois Archaeological Survey. Dickson Mounds Museum, Lewistown.

1996 Excavations at Iliniwek Village State Historic Site: The Illini Village of the Jolliet/Marquette Expedition of 1673. Paper presented at the 29th Annual SHA Conference on Historical and Underwater Archaeology, Cincinnati.

Grantham, Larry D., and Kathleen L. Ehrhardt

2000 An Evaluation of Danner Ware as Seen from the Excavations at the Iliniwek Village State Historic Site, Clark County, Missouri. Paper presented at the Joint Midwest Archaeological and Plains Anthropological Conference, St. Paul.

Greber, N'omi, and Katharine C. Ruhl

1989 *The Hopewell Site: A Contemporary Analysis Based on the Work of Charles C. Willoughby.* Westview Press, Boulder.

Griffin, James B. (editor)

1961 *Lake Superior Copper and the Indians: Miscellaneous Studies of Great Lakes Prehistory.* Anthropological Papers No. 17. Museum of Anthropology, University of Michigan, Ann Arbor.

Hall, Robert L.

1991 The Archaeology of La Salle's Fort St. Louis on Starved Rock and the Problem of the Newell Fort. In *French Colonial Archaeology: The Illinois Country and the Western Great Lakes,* edited by John A. Walthall, pp. 14–28. University of Illinois Press, Urbana.

1993 Red Banks, Oneota, and the Winnebago: Views from a Distant Rock. *Wisconsin Archeologist* 74(1–4):10–79.

Halsey, John

1992 *Miskwabik—Red Metal: The Roles Played by Michigan's Copper in Prehistoric North America.* Reprinted. Keweenaw County Historical Society. Originally published 1983, Michigan History Magazine, September/October.

1996 Without Forge or Crucible: Aboriginal Native American Use of Metals and Metallic Ores in the Eastern Woodlands. *Michigan Archaeologist* 42(1):1–58.

Hamell, George R.

1983 Trading in Metaphors: The Magic of Beads: Another Perspective upon Indian-European Contact in Northeastern North America. In *Proceedings of the 1982 Glass Bead Conference,* edited by Charles F. Hayes III, pp. 5–28. Research Records No. 16. Rochester Museum and Science Center, Rochester.

1986 Strawberries, Floating Islands, and Rabbit Captains: Mythical Realities and European Contact in the Northeast during the Sixteenth and Seventeenth Centuries. *Journal of Canadian Studies* 21:72–94.

Hamilton, Elizabeth G.

1996 *Technology and Social Change in Belgic Gaul: Copper Working at the Titelberg, Luxembourg, 125 B.C.–A.D. 300.* Research Papers in Science and Archaeology, Vol. 13. MASCA, University Museum of Archaeology and Anthropology, University of Pennsylvania, Philadelphia.

Hamilton, Henry W., Jean Tyree Hamilton, and Eleanor F. Chapman

1974 *Spiro Mound Copper.* Memoir No.11. Missouri Archaeological Society, Columbia.

Hancock, R. G. V.

2000 Elemental Analysis. In *Modern Analytical Methods in Art and Archaeology,* edited by Enrico Ciliberto and Giuseppe Spoto, pp. 11–20. John Wiley and Sons, New York.

Hancock, R. G. V., R. M. Farquhar, L. A. Pavlish, and W. D. Finlayson

1995 Chemical Analysis of Copper and Brass Samples from Christian Island, Georgian Bay, Ontario. *Journal of Radioanalytical and Nuclear Chemistry, Articles* 196(2):255–266.

Hancock, R. G. V., W. A. Fox, T. Conway, and L. A. Pavlish

1993 Chemical Analysis of Archaeological Copper and Brass from Northeastern Ontario. *Journal of Radioanalytical and Nuclear Chemistry* 168(2):307–315.

Hancock, R. G. V., L. A. Pavlish, R. M. Farquhar, P. J. Julig, and W. A. Fox
1994 Chemical Seriation of Northeastern North American Archaeological Sites Using Copper Artifacts. In *Archaeometry of Pre-Columbian Sites and Artifacts,* edited by David A. Scott and Pieter Meyers, pp. 255–265. Getty Conservation Institute, Los Angeles.

Hancock, R. G. V., L. A. Pavlish, R. M. Farquhar, R. Salloum, W. A. Fox, and G. C. Wilson
1991 Distinguishing European Trade Copper and North-Eastern North American Native Copper. *Archaeometry* 33(1):69–86.

Hancock, R. G. V., L. A. Pavlish, W. A. Fox, and M. A. Latta
1995 Chemical Analysis of Copper Alloy Trade Metal from a Post-Contact Huron Site in Ontario, Canada. *Archaeometry* 37(2):339–350.

Hauser, Raymond E.
1973 *An Ethno-History of the Illinois Indian Tribe, 1673–1832.* Unpublished Ph.D. dissertation, Northern Illinois University, DeKalb. University Microfilms, Ann Arbor.

Hegmon, Michelle
1998 Technology, Style, and Social Practices: Archaeological Approaches. In *The Archaeology of Social Boundaries,* edited by Miriam Stark, pp. 264–279. Smithsonian Institution Press, Washington, D.C.

Helms, Mary W.
1993 *Craft and the Kingly Ideal: Art, Trade, and Power.* University of Texas Press, Austin.

Henning, Dale
1970 Development and Interrelationships of Oneota Culture in the Lower Missouri River Valley. *Missouri Archaeologist* 32:1–80.

Herbert, Eugenia
1984 *Red Gold of Africa: Copper in Precolonial History and Culture.* University of Wisconsin Press, Madison.

Hickerson, Harold
1971 The Chippewa of the Upper Great Lakes. In *North American Indians in Historical Perspective,* edited by Eleanor Burke Leacock and Nancy Oestreich Lurie, pp. 169–199. Waveland Press, Prospect Heights.

Hodder, Ian
1986 *Reading the Past: Current Approaches to Interpretation in Archaeology.* Cambridge University Press, Cambridge.

Hodges, Henry
1976 *Artifacts: An Introduction to Early Materials and Technology.* John Baker, London.

Hoffman, Christopher R., and Marcia-Anne Dobres
1999 Conclusion: Making Material Culture, Making Culture Material. In *The So-*

cial Dynamics of Technology, edited by Marcia-Anne Dobres and Christopher R. Hoffman, pp. 209–222. Smithsonian Institution Press, Washington, D.C.

Holl, Augustin F. C.

2000 Metals and Precolonial African Society. In *Ancient African Metallurgy,* edited by Joseph O. Vogel, pp. 1–81. Altamira Press, Walnut Creek, California.

Hollinger, R. Eric

2001 *Report on the Iliniwek Village State Historic Site (23CK116) Paleoethnobotanical Analysis.* Report submitted to the Missouri Division of State Parks, Department of Natural Resources. Ms. on file, Missouri Department of Natural Resources, Jefferson City.

Hosler, Dorothy

1994 *The Sounds and Colors of Power: The Sacred Metallurgical Technology of Ancient West Mexico.* MIT Press, Cambridge.

Hughes, Thomas

1987 The Seamless Web: Technology, Science, et Cetera, et Cetera. In *Technology and Social Process,* edited by Brian Elliot, pp. 9–19. Edinburgh University Press, Edinburgh.

1990 From Deterministic Dynamos to Seamless-Web Systems. In *Engineering as a Social Enterprise,* edited by H. Sladovich, pp. 7–25. National Academic Press, Washington, D.C.

Hungry Wolf, Adolf

1999 *Pow-Wow: Dancer's and Craftworker's Handbook.* Canadian Caboose Press, Skookumchuck, British Columbia.

Jablow, Joseph

1974 *Indians of Illinois and Indiana.* Garland Publishing, New York.

Jaenen, Cornelius

1974 Amerindian Views of French Culture in the Seventeenth Century. *Canadian Historical Review* LV(3):271–291.

1996 *The French Regime in the Upper Country of Canada during the Seventeenth Century.* Champlain Society, Toronto.

Karklins, Karlis

1992 *Trade Ornament Usage among the Native Peoples of Canada: A Source Book.* National Historic Sites Parks Service, Environment Canada, Ottawa.

Kehoe, Alice Beck

1981 *North American Indians, a Comprehensive Account.* Prentice-Hall, Englewood Cliffs, New Jersey.

Kellogg, Louise Phelps

1917 *Early Narratives of the Northwest, 1634–1699.* Charles Scribner's Sons, New York.

1968 *The French Régime in Wisconsin and the Northwest.* Cooper Square Publishers, New York.

Kent, Barry

1984 *Susquehanna's Indians.* Pennsylvania Historical and Museum Commission, Harrisburg.

Kenyon, W. A.

1982 *The Grimsby Site: A Historic Neutral Cemetery.* Royal Ontario Museum, Toronto.

Kidd, Kenneth E.

1953 The Excavation and Historical Identification of a Huron Ossuary. *American Antiquity* 18(4):359–379.

Kingery, W. David

1993 Technological Systems and Some Implications with Regard to Continuity and Change. In *History from Things: Essays on Material Culture,* edited by Steven Lubar and W. David Kingery, pp. 215–230. Smithsonian Institution Press, Washington, D.C.

1996a Introduction. In *Learning from Things: Method and Theory of Material Culture Studies,* edited by W. David Kingery, pp. 1–15. Smithsonian Institution Press, Washington, D.C.

1996b Materials Science and Material Culture. In *Learning from Things: Method and Theory of Material Culture Studies,* edited by W. David Kingery, pp. 181–203. Smithsonian Institution Press, Washington, D.C.

Kinietz, W. Vernon

1996 *The Indians of the Western Great Lakes, 1615–1760.* Reprinted. Ann Arbor Paperbacks. Originally published 1940, University of Michigan Press, Ann Arbor.

Kopytoff, Igor

1986 The Cultural Biography of Things: Commoditization as Process. In *The Social Life of Things: Commodities in Cultural Perspective,* edited by Arjun Appadurai, pp. 64–91. Cambridge University Press, Cambridge.

Kraft, Herbert C.

1996 *The Dutch, the Indians, and the Quest for Copper: Pahaquarry and the Old Mine Road.* Seton Hall University Museum, South Orange.

Kuleff, I., and E. Pernicka

1995 Instrumental Neutron Activation Analysis of Native Copper: Some Methodological Considerations. *Journal of Radioanalytical and Nuclear Chemistry, Articles* 191(1):145–161.

Kupp, Jan

1968 Fur Trade Relations, New Netherland—New France, 1600–1664. Unpublished Ph.D. dissertation, University of Manitoba.

Lascoe, O. D.

1988 *Handbook of Fabrication Processes.* American Society for Metals International, Materials Park, Ohio.

Latta, Martha A., and Lisa Anselmi

1995 Expediency and Curation of Metal Tools: The Technology of Cuprous "Scrap" at the Thomson-Walker and Auger Sites in Southern Ontario. Paper presented at the 41st Midwest Archaeological Conference, Beloit.

Latta, Martha A., Paul Thibaudeau, and Lisa Anselmi

1998 Expediency and Curation: The Use and Distribution of "Scrap" Trade Metal by Huron Native Peoples in Sixteenth-Century Southern Ontario. *Wisconsin Archeologist* 79(1):175–184.

Leader, Jonathan

1988 Technological Continuities and Specialization in Prehistoric Metalwork in the Eastern United States. Unpublished Ph.D. dissertation, University of Florida. University Microfilms, Ann Arbor.

Lechtman, Heather

1977 Style in Technology—Some Early Thoughts. In *Material Culture: Styles, Organization, and Dynamics of Technology,* edited by Heather Lechtman and R. Merrill, pp. 3–20. West Publishing Company, St. Paul.

1994 The Materials Science of Material Culture: Examples from the Andean Past. In *Archaeometry of Pre-Columbian Sites and Artifacts,* edited by David A. Scott and Pieter Meyers, pp.3–12. Getty Conservation Institute, Los Angeles.

Lemonnier, Pierre

1980 Les Salines de l'ouest—logique technique, logique sociale. Maison des Sciences de l'Homme/Presses Universitaires de Lille. Paris/Lille.

1986 The Study of Material Culture Today: Toward an Anthropology of Technical Systems. *Journal of Anthropological Archaeology* 5(2):147–186.

1993 Introduction. In *Technological Choices: Transformation in Material Cultures since the Neolithic,* edited by Pierre Lemonnier, pp. 1–35. Routledge, London.

Leone, Mark P., and Parker B. Potter Jr.

1988 Introduction: Issues in Historical Archaeology. In *The Recovery of Meaning: Historical Archaeology in the Eastern United States,* edited by Mark P. Leone and Parker B. Potter Jr., pp. 1–11. Smithsonian Institution Press, Washington, D.C.

Leute, Ulrich

1987 *Archaeometry: An Introduction to Physical Methods in Archaeology and the History of Art.* VCH Verlagsgesellschaft, Weinheim.

Levine, Mary Ann

1996 *Native Copper, Hunter-Gatherers, and Northeastern Prehistory.* Unpublished Ph.D. dissertation, University of Massachusetts, Amherst. University Microfilms, Ann Arbor.

1999 Native Copper in the Northeast: An Overview of Potential Sources Available to Indigenous Peoples. In *The Archaeological Northeast,* edited by Mary

Ann Levine, Kenneth E. Sassaman, and Michael S. Nassaney, pp. 183–199. Bergin and Garvey, Westport, Connecticut.

Lightfoot, Kent

1995 Culture Contact Studies: Redefining the Relationship between Prehistoric and Historical Archaeology. *American Antiquity* 60(2):199–217.

Lindsey, E. M.

1964 Cooper Farm Salvage Project. *Journal of Alabama Archaeology* X(1):22–29.

Little, Barbara (editor)

1992 *Text-Aided Archaeology.* CRC Press, Boca Raton, Florida.

Loney, Helen

2000 A Critical Review of Models of Technological Change in Ceramic Studies. *American Antiquity* 65(4):646–668.

Longoria, Linda

1998 Surface Artifacts from the Iliniwek Village Site, Northeast Missouri. *Missouri Archaeologist* 59:125–151.

Lurie, Rochelle

1998 The Late Prehistoric and Proto-Historic Components at the New Lenox Site in Will County, Illinois: A Coalescence of Peoples in a Time of Trouble? Paper presented at the Museums at Prophetstown Woodland National Conference, Indiana Dunes National Lakeshore, Porter, Indiana.

Lyman, Taylor (editor)

1972 *Atlas of Microstructures of Industrial Alloys.* Metals Handbook, Vol. 7. 8th edition. American Society for Metals, Materials Park.

Maceachern, Scott

1998 Scale, Style, and Cultural Variation: Technological Traditions in the Northern Mandara Mountains. In *The Archaeology of Social Boundaries,* edited by Miriam Stark, pp.107–131. Smithsonian Institution Press, Washington, D.C.

Maddin, R. T., Stech Wheeler, and J. D. Muhly

1980 Distinguishing Artifacts Made of Native Copper. *Journal of Archaeological Science* 7:211–225.

Malone, Patrick

1991 *The Skulking Way of War: Technology and Tactics among the New England Indians.* Johns Hopkins University Press, Baltimore.

Margry, Pierre (compiler and editor)

1879–1888 *Découvertes et établissements des Français dans l'ouest et dans le sud de l'Amérique septentrionale (1614–1754): Mémoires et documents originaux.* 6 vols. Maisonneuve, Paris.

Martin, Calvin

1975 The Four Lives of a Micmac Copper Pot. *Ethnohistory* 22(2):111–129.

Martin, Susan

1999 *Wonderful Power: The Story of Ancient Copper Working in the Lake Superior Basin.* Wayne State University Press, Detroit.

Martin, Terrance J., J. Chris Richmond, and Erin L. Brand

2003 *An Archaeozoological Analysis of 23CK116, the Illini Village of the Marquette and Joliet Voyage of 1673.* Illinois State Museum Landscape History Program Technical Report 2003–1478–8 (Springfield). Report prepared for the Missouri Department of Natural History, Jefferson City.

Mason, Carol I.

1976 Historic Identification and Lake Winnebago Focus Oneota. In *Culture Change and Continuity: Essays in Honor of James Bennett Griffin,* edited by Charles E. Cleland, pp. 335–367. Academic Press, New York.

1997 The Historic Period in Wisconsin Archaeology, Native Peoples. *Wisconsin Archeologist* 78(1–2):298–319.

Mason, Ronald

1976 Ethnicity and Archaeology in the Upper Great Lakes. In *Cultural Change and Continuity: Essays in Honor of James Bennett Griffin,* edited by Charles E. Cleland, pp. 349–361. Academic Press, New York.

1981 *Great Lakes Archaeology.* Academic Press, New York.

1986 Rock Island: Historical Indian Archaeology in the Northern Lake Michigan Basin. Midcontinental Journal of Archaeology (MCJA) Special Paper No. 6. Kent State University Press, Kent.

Masthay, Carl (editor)

2002 *Kaskaskia Illinois-to-French Dictionary.* Published by the author, St. Louis.

McCreight, Tim

1991 *The Complete Metalsmith: An Illustrated Handbook.* Davis Publications, Worcester, Massachusetts.

McEwan, Bonnie G. (editor)

2000 *Indians of the Greater Southeast: Historical Archaeology and Ethnohistory.* University Press of Florida, Gainesville.

Merrell, James H.

1988 The Indians' New World: The Catawba Experience. In *Material Life in America, 1600–1860,* edited by Robert Blair St. George, pp. 95–112. Northeastern University Press, Boston.

Milanich, Jerald J.

1999 *Laboring in the Fields of the Lord.* Smithsonian Institution Press, Washington, D.C.

Miller, Christopher, and George Hamell

1986 A New Perspective on Indian-White Contact: Cultural Symbols and Colonial Trade. *Journal of American History* 73:311–328.

Moore, Clarence B.
1903 Sheet-Copper from the Mounds is Not Necessarily of European Origin. *American Anthropologist* 5(1):27–49.

Morantz, Toby
1992 Old Texts, Old Questions: Another Look at the Issue of Continuity and the Early Fur Trade Period. *Canadian Historical Review* 73(2):166–193.

Moreau, J. F., and R. G. V. Hancock
1999 Faces of European Copper Alloy Cauldrons from Québec and Ontario "Contact" Sites. In *Metals in Antiquity,* edited by Suzanne M. M. Young, A. Mark Pollard, Paul Budd, and Robert A. Ixer, pp.326–333. BAR International Series 792. Archaeopress, Oxford.

Morrell, L. Ross
1965 *The Woods Island Site in Southeastern Acculturation, 1625–1800.* Florida State University Notes in Anthropology 11, Tallahassee.

Morse, Dan F.
1992 The Seventeenth-Century Michigamea Village Location in Arkansas. In *Calumet and Fleur-de-Lys: Archaeology of Indian and French Contact in the Midcontinent,* edited by John A. Walthall and Thomas E. Emerson, pp. 55–74. Smithsonian Institution Press, Washington, D.C.

Morse, Dan F., and Phyllis A. Morse
1983 *Archaeology of the Central Mississippi Valley.* Academic Press, San Diego.

Moussette, Marcel
2001 Les médailles religiueses, une forme de l'imagerie baroque en Nouvelle-France. *Les Cahiers des Dix* 55:295–329.

Muller, Jon
1989 The Southern Cult. In *The Southeastern Ceremonial Complex: Artifacts and Analysis,* edited by Patricia Galloway, pp. 11–37. University of Nebraska Press, Lincoln.

Nassaney, Michael S., and Eric S. Johnson
2000 Contributions of Material Objects to Ethnohistory in Native North America. In *Interpretations of Native North American Life,* edited by Michael S. Nassaney and Eric S. Johnson, pp. 1–30. University Press of Florida, Gainesville.

Neff, Hector
1994 RQ-Mode Principal Components Analysis of Ceramic Compositional Data. Archaeometry 36(1):115–130.
2000 Neutron Activation Analysis for Provenance Determination in Archaeology. In *Modern Analytical Methods in Art and Archaeology,* edited by Enrico Ciliberto and Giuseppe Spoto, pp. 81–134. John Wiley and Sons, New York.

Neff, H., and M. Glascock
1995 The State of Nuclear Archaeology in North America. *Journal of Radioanalytical and Nuclear Chemistry, Articles* 196(2):275–286.

Neill, Rev. Edward D.

1884 Discovery along the Great Lakes. In *Narrative and Critical History of America*, Vol. 4, edited by Justin Winsor, pp. 163–199. Houghton, Mifflin and Company, The Riverside Press, Cambridge.

Newman, Roger C., and Karl Sieradzki

1994 Metallic Corrosion. *Science* 263:1708–1709.

Ogden, Jack

1991 Classical Gold Wire: Some Aspects of its Manufacture and Use. *Jewellery Studies* 5:95–105.

Paterek, Josephine

1994 *Encyclopedia of American Indian Costume.* W. W. Norton, New York.

Pease, Theodore Calvin, and Raymond C. Werner (editors)

1934 Memoir of De Gannes Concerning the Illinois Country. In *The French Foundations, 1680–1693.* Collections of the Illinois State Historical Library 23:302–395, Springfield.

Penman, John T.

1977 The Old Copper Culture: An Analysis of Old Copper Artifacts. *Wisconsin Archeologist* 58(1):3–23.

Penney, David

1985 Continuities of Imagery and Symbolism in the Art of the Woodlands. In *Ancient Art of the American Woodland Indians,* edited by D. Brose, J. Brown, and D. Penney, pp. 147–231. Harry N. Abrams, New York.

Perrin, Noel

1979 *Giving Up the Gun: Japan's Reversion to the Sword, 1543–1879.* David R. Godine, Boston.

Perttula, Timothy K.

1991 European Contact and Its Effects on Aboriginal Caddoan Populations between A.D. 1520 and A.D. 1680. In *The Spanish Borderlands in Pan-American Perspective,* edited by David Hurst Thomas, pp. 501–518. Columbian Consequences, Vol. 3. Smithsonian Institution Press, Washington, D.C.

Pfaffenberger, Bryan

1992 Social Anthropology of Technology. *Annual Review of Anthropology* 21: 491–516.

2001 Symbols Do Not Create Meanings—Activities Do: Or, Why Symbolic Anthropology Needs the Anthropology of Technology. In *Anthropological Perspectives on Technology,* edited by Michael Brian Schiffer, pp. 77–86. University of New Mexico Press, Albuquerque.

Pigott, Vincent C.

1991 Discussion of *Metals in Society.* In *Metals in Society: Theory beyond Analysis,* edited by Robert M. Ehrenreich, pp. 81–84. Research Papers in Science and

Archaeology Vol. 8, Part 2. MASCA, University Museum of Archaeology and Anthropology, University of Pennsylvania, Philadelphia.

Pollard, A. Mark, and Carl Heron

1996 *Archaeological Chemistry.* The Royal Society of Chemistry, Cambridge.

Quimby, George

1966 *Indian Culture and European Trade Goods.* University of Wisconsin Press, Madison.

Quimby, George I., and Alexander Spoehr

1951 Acculturation and Material Culture—I. *Fieldiana: Anthropology* 36:107–147.

Ramenofsky, Ann F.

1991 Historical Science and Contact Period Studies. In *The Spanish Borderlands in Pan-American Perspective,* edited by David Hurst Thomas, pp. 437–452. Columbian Consequences, Vol. 3. Smithsonian Institution Press, Washington, D.C.

1998 Evolutionary Theory and the Native American Record of Artifact Replacement. In *Studies in Culture Contact: Interaction, Culture Change, and Archaeology,* edited by James G. Cusick, pp. 77–101. Occasional Paper No. 25. Center for Archaeological Investigations, Southern Illinois University, Carbondale.

Rapp, George (Rip), James Allert, Vanda Vitali, Zhichun Jing, and Eiler Henrickson

2000 *Determining Geologic Sources of Artifact Copper: Source Characterization Using Trace Element Patterns.* University Press of America, Lanham.

Reber, Eleanora A.

1997 Protohistoric Fort Ancient Copper Artifacts from the Madisonville Site: A Metallographic Examination. Preliminary Report.

Redfield, Robert, Ralph Linton, and Melville J. Herskovits

1936 Memorandum for the Study of Acculturation. *American Anthropologist* 38:149–152.

Redman, Charles R.

1975 Productive Sampling Strategies for Archaeological Sites. In *Sampling in Archaeology,* edited by James W. Mueller, pp. 147–154. University of Arizona Press, Tucson.

Rogers, J. Daniel

1990 *Objects of Change.* Smithsonian Institution Press, Washington, D.C.

1993 The Social and Material Implications of Culture Contact on the Northern Plains. In *Ethnohistory and Archaeology: Approaches to Postcontact Change in the Americas,* edited by J. Daniel Rogers and Samuel M. Wilson, pp. 73–88. Plenum Press, New York.

Roundtree, Helen C.

1989 *The Powhatan Indians of Virginia.* University of Oklahoma Press, Norman.

Rubertone, Patricia E.

1989 Archaeology, Colonialism, and 17th Century Native America: Towards an Alternative Interpretation. In *Conflict in the Archaeology of Living Traditions,* edited by Robert Layton, pp. 32–45. Routledge, London.

2001 *Grave Undertakings: An Archaeology of Roger Williams and the Narragansett Indians.* Smithsonian Institution Press, Washington, D.C.

Ruhl, Katharine C.

1992 Copper Ear Spools from Ohio Hopewell Sites. *Midcontinental Journal of Archaeology* 17(1):46–72.

Ruhl, Katharine C., and M. F. Seeman

1998 Temporal and Social Implications of Ohio Hopewell Copper Ear Spool Design. *American Antiquity* 63(4):651–662.

Sabo III, George

2000 The Quapaw Indians of Arkansas, 1673–1803. In *Indians of the Greater Southeast: Historical Archaeology and Ethnohistory,* edited by Bonnie G. McEwan, pp. 178–203. University Press of Florida, Gainesville.

Sahlins, Marshall

1981 *Historical Metaphors and Mythical Realities: Structure in the Early History of the Sandwich Island Kingdom.* Association for Social Anthropology in Oceania Special Papers No. 1. University of Michigan Press, Ann Arbor.

Samans, Carl H.

1963 *Metallic Materials in Engineering.* Macmillan, New York.

Sampson, Kelvin, and Duane Esarey

1993 A Survey of Elaborate Mississippian Copper Artifacts from Illinois. In *Highways to the Past: Essays on Illinois Archaeology in Honor of Charles J. Bareis,* edited by Thomas E. Emerson, Andrew C. Fortier, and Dale L. McElrath, pp. 452–480. *Illinois Archaeology* 5(1–2):452–480.

Saunders, Rebecca

1998 Forced Relocation, Power Relations, and Culture Contact in the Missions of La Florida. In *Studies in Culture Contact: Interaction, Culture Change, and Archaeology,* edited by James G. Cusick, pp. 402–429. Occasional Paper No. 25. Center for Archaeological Investigations, Southern Illinois University, Carbondale.

Sayre, Gordon M.

1997 *Les Sauvages Américains: Representations of Native Americans in French and English Colonial Literature.* University of North Carolina Press, Chapel Hill.

Schaniel, William

1988 New Technology and Culture Change in Traditional Societies. *Journal of Economic Issues* 22:493–498.

Schiffer, Michael Brian

2001 Toward an Anthropology of Technology. In *Anthropological Perspectives on Technology,* ed. by Schiffer, pp. 1–15. University of New Mexico Press, Albuquerque.

Schlanger, Nathan

1994 Mindful Technology: Unleashing the Chaîne Opératoire for an Archaeology of Mind. In *The Ancient Mind: Elements of Cognitive Archaeology,* edited by Colin Renfrew and Ezra B. W. Zubrow, pp. 143–151. Cambridge University Press, Cambridge.

Schnell, Gail Schroeder

1974 *Hotel Plaza.* Reports of Investigations No. 29. Illinois State Museum, Springfield.

Schortman, Edward M., and Patricia A. Urban

1998 Culture Contact Structure and Process. In *Studies in Culture Contact: Interaction, Culture Change, and Archaeology,* edited by James G. Cusick, pp. 102–125. Occasional Paper No. 25. Center for Archaeological Investigations, Southern Illinois University, Carbondale.

Schroeder, David L., and Katharine C. Ruhl

1968 Metallurgical Characteristics of North American Prehistoric Copper Work. *American Antiquity* 33(2):162–169.

Scott, David A.

1991 *Metallography and Microstructure of Ancient and Historic Metals.* Getty Conservation Institute and the J. Paul Getty Museum, Los Angeles.

Seed, Patricia

1995 *Ceremonies of Possession in Europe's Conquest of the New World, 1492–1640.* Cambridge University Press, Cambridge.

Seeman, Mark F.

1979 *The Hopewell Interaction Sphere: The Evidence for Interregional Trade and Structural Complexity.* Prehistory Research Series, Vol. 2. Indiana Historical Society, Indianapolis.

Sempowski, Martha L., and Lorraine R. Saunders

2001 Dutch Hollow and Factory Hollow: The Advent of Dutch Trade among the Seneca. Charles F. Wray Series in Seneca Archaeology, Vol. 3. Research Records 24. Rochester Museum and Science Center, Rochester.

Sigaut, François

1994 Technology. In *The Companion Encyclopedia of Anthropology,* edited by Tim Ingold, pp. 420–459. Routledge, London.

Smith, Cyril Stanley

1968 Metallographic Study of Early Artifacts Made from Native Copper. In *Actes du XIe Congres International d'Histoire des Sciences VI,* pp. 237–252. Warsaw.

1981a The Interpretation of Microstructures of Metallic Artifacts. In *A Search for Structure: Selected Essays on Science, Art, and History,* pp. 69–111. MIT Press, Cambridge. Originally published 1967, in *Application of Science in Examination of Works of Art,* edited by W. J. Young, pp. 20–52. Museum of Fine Arts, Boston.

1981b Structural Hierarchy in Science, Art, and History. In *A Search for Structure: Selected Essays on Science, Art, and History,* pp. 358–390. MIT Press, Cambridge. Originally published 1978, in *On Aesthetics in Science,* edited by Judith Wechsler, pp. 9–53. MIT Press, Cambridge.

Smith, Marvin T.

1987 *Archaeology of Aboriginal Culture Change in the Interior Southeast: Depopulation during the Early Historic Period.* University Press of Florida/Florida Museum of Natural History, Gainesville.

Smith, Merritt Roe

1985 Social Processes and Technological Change. *Reviews in American History,* June:157–166.

South, Stanley (editor)

1977 *Research Strategies in Historical Archaeology.* Academic Press, New York.

Spicer, Edward

1961 Types of Contact and Processes of Change. In Perspectives in American Indian Culture Change, ed. by Spicer, pp. 517–544. University of Chicago Press, Chicago.

Spier, Robert

1968 Technology and Material Culture. In *Introduction to Cultural Anthropology: Essays in the Scope and Methods of the Science of Man,* edited by James A. Clifton, pp. 131–159. Houghton Mifflin, Boston.

Stark, Miriam (editor)

1998 *The Archaeology of Social Boundaries.* Smithsonian Institution Press, Washington, D.C.

Stevenson, Christopher

1976 Patterns on Old Copper Artifacts. *Wisconsin Archeologist* 57(2):138–139.

Stone, Lyle M., and Donald Chaput

1978 History of the Upper Great Lakes Area. In *Northeast,* edited by Bruce G. Trigger, pp. 602–609. Handbook of North American Indians, Vol. 15, William C. Sturtevant, general editor, Smithsonian Institution, Washington, D.C.

Stothers, David M.

1991 Beads, Brass, and Beaver: Archaeological Reflections of Protohistoric "Fire Nation" Trade and Exchange. *Archaeology of Eastern North America* 19: 121–134.

2000 The Protohistoric Time Period in the Southwestern Lake Erie Region: European-Derived Trade Material, Population Movement, and Cultural Realignment. In *Cultures before Contact: The Late Prehistory of Ohio and Surrounding Regions,* edited by Robert A. Genheimer, pp. 52–94. Ohio Archaeological Council, Columbus.

Struever, Stuart, and Gail. L. Houart

1972 An Analysis of the Hopewell Interaction Sphere. In *Social Exchange and Interaction,* edited by E. Wilmsen, pp. 47–79. Anthropological Papers No. 46. Museum of Anthropology, University of Michigan, Ann Arbor.

Tanner, Helen Hornbeck (editor)

1987 *Atlas of Great Lakes Indian History.* University of Oklahoma Press, Norman.

Temple, Wayne C.

1966 *Indian Villages of the Illinois Country.* Scientific Papers, Vol. II, Part 2. Illinois State Museum, Springfield.

Thomas, David Hurst

1989 Columbian Consequences, the Spanish Borderlands in Cubist Perspective. In *Archaeological and Historical Perspectives on the Spanish Borderlands West,* edited by David Hurst Thomas, pp. 1–14. Columbian Consequences, Vol. 1. Smithsonian Institution Press, Washington, D.C.

1991 Cubist Perspectives on the Spanish Borderlands: Past, Present, and Future. In *The Spanish Borderlands in Pan-American Perspective,* edited by David Hurst Thomas, pp. xv–xxi. Columbian Consequences, Vol. 3. Smithsonian Institution Press, Washington, D.C.

Thomas, Nicholas

1991 *Entangled Objects.* Harvard University Press, Cambridge.

Thwaites, Reuben Gold

1902 *Father Marquette.* D. Appleton and Co., New York.

Thwaites, Reuben Gold (editor)

1896–1901 *The Jesuit Relations and Allied Documents.* 73 vols. Burrows Brothers Company, Cleveland.

1903 *Father Louis Hennepin: A New Discovery of a Vast Country in America,* Vol. 1. A. C. McClurg and Co., Chicago.

Tiffany, Joseph

1988 Preliminary Report on Excavations at the McKinney Oneota Village Site (13LA1), Louisa County, Iowa. *Wisconsin Archeologist* 69:228–312.

Trigger, Bruce

1980 Archaeology and the Image of the American Indian. *American Antiquity* 45:662–76.

1985 *Natives and Newcomers: Canada's "Heroic Age" Reconsidered.* McGill-Queen's University Press, Kingston, Ontario.

1986 Ethnohistory: The Unfinished Edifice. *Ethnohistory* 41(3):369–405.

1987 *The Children of Aataentsic: A History of the Huron People to 1660.* 2 vols. Reprinted. McGill-Queen's University Press, Montreal and Kingston. Originally published 1976, McGill-Queen's University Press, Montreal.

1989 History and Contemporary Archaeology. In *Archaeological Thought in America,* edited by C. C. Lamberg-Karlovsky, pp. 19–34. Cambridge University Press, Cambridge.

1991 Early Native North American Responses to European Contact: Romantic Versus Rationalistic Interpretations. *Journal of American History,* March: 1195–1215.

Turgeon, Laurier

1997 The Tale of the Kettle: Odyssey of an Intercultural Object. *Ethnohistory* 44(1):1–29.

Untracht, Oppi

1968 *Metal Techniques for Craftsmen.* Doubleday, Garden City, New York.

Vander Voort, George F.

1999 *Metallography: Principles and Practice.* American Society for Metals International, Materials Park, Ohio.

Vernon, W. W.

1985 New Perspectives on the Archaeometallurgy of the Old Copper Industry. *MASCA Journal* 3(5):154–163.

1990 New Archaeometallurgical Perspectives on the Old Copper Industry of North America. In *Archaeological Geology of North America,* edited by N. P. Lasca and J. Donahue, pp. 499–512. Centennial Special Volume 4, Geological Society of America, Boulder.

Wagner, Mark J.

1998 Some Think It Impossible to Civilize Them at All: Cultural Change and Continuity among the Early Nineteenth-Century Potawatomi. In *Studies in Culture Contact: Interaction, Culture Change, and Archaeology,* edited by James G. Cusick, pp. 430–456. Occasional Paper No. 25. Center for Archaeological Investigations, Southern Illinois University, Carbondale.

Walker, C., R. G. V. Hancock, S. Aufreiter, M. A. Latta, and C. Garrad

1999 Chronological Markers? Chemical Analysis of Copper-Based Trade Metal Artefacts from Petun Sites in Southern Ontario, Canada. In *Metals in Antiquity,* edited by Suzanne M. M. Young, A. Mark Pollard, Paul Budd, and Robert A. Ixer, pp. 317–325. BAR International Series 792. Archaeopress, Oxford.

Walters, Gary Rex, and Roger Boyd

1986 Phase I: Cultural Resources Survey and Evaluation of Parts of Township 65 North, Range 6 West, Clark County, for the Missouri Department of Natural Resources, Historic Preservation Program, Division of Parks and Recreation, Jefferson City, Missouri. On file, Missouri Department of Natural Resources, Jefferson City.

Walthall, John A.

1992 Aboriginal Pottery and the Eighteenth-Century Illini. In *Calumet and Fleur-de-Lys: Archaeology of Indian and French Contact in the Midcontinent,* edited by John A. Walthall and Thomas E. Emerson, pp. 155–174. Smithsonian Institution Press, Washington, D.C.

Walthall, John A., and Elizabeth D. Benchley

1987 *The River L'Abbe Mission.* Studies in Illinois Archaeology No. 2. Illinois Historic Preservation Agency, Springfield.

Walthall, John A., and Margaret Kimball Brown

2001 French Colonial Material Culture from an Early Eighteenth-Century Outpost in the Illinois Country. *Illinois Archaeology* 13(1 and 2):88–126.

Walthall, John A., and Thomas E. Emerson

1991 French Colonial Archaeology. In *French Colonial Archaeology: The Illinois Country and the Western Great Lakes,* edited by John A. Walthall, pp. 1–13. University of Illinois Press, Urbana.

Walthall, John A., F. Terry Norris, and Barbara D. Stafford

1992 Woman Chief's Village: An Illini Winter Hunting Camp. In *Calumet and Fleur-de-Lys: Archaeology of Indian and French Contact in the Midcontinent,* edited by John A. Walthall and Thomas E. Emerson, pp. 129–153. Smithsonian Institution Press, Washington, D.C.

Walz, Gregory R.

1996 Late Prehistoric and Proto-Historic Subsistence at the Zimmerman Site: The Archaeobotanical Evidence. Paper presented at the 29th Annual SHA Conference on Historical and Underwater Archaeology, Cincinnati.

Washburn, Wilcomb E.

1967 Symbol, Utility, and Aesthetics in the Indian Fur Trade. In *Aspects of the Fur Trade: Selected Papers of the 1965 North American Fur Trade Conference,* edited by R. W. Fridley, pp. 50–54. Minnesota Historical Society, St. Paul.

Wayman, Michael L.

1989 Native Copper: Humanity's Introduction to Metallurgy? In *All That Glitters: Readings in Historical Metallurgy,* edited by Michael L. Wayman, pp. 3–12. Metallurgical Society of the Canadian Institute of Mining and Metallurgy, Montreal.

2000 Archaeometallurgical Contributions to a Better Understanding of the Past. *Materials Characterization* 45(4/5):259–267.

Wayman, Michael L., and M. John M. Duke

1999 The Effects of Melting on Native Copper. *Der Anschnitt, Beilheft* 9:55–63.

Wayman, Michael L., J. C. H. King, and P. T. Craddock

1992 *Aspects of Early North American Metallurgy.* British Museum Occasional Paper 79. Departments of Scientific Research and Ethnography, London.

Wayman, Michael L., R. R. Smith, C. G. Hickey, and M. J. M. Duke

1985 The Analysis of Copper Artefacts of the Copper Inuit. *Journal of Archaeological Science* 12:367–375.

Wedel, Mildred Mott

1959 Oneota Sites on the Upper Iowa River. *Missouri Archaeologist* 21(2–4):1–181.

Wells, Peter

1991 Metals and Meaning. In *Metals in Society: Theory beyond Analysis,* edited by Robert M. Ehrenreich, pp. 89–92. Research Papers in Science and Archaeology, Vol. 8, Part 2. MASCA, University Museum of Archaeology and Anthropology, University of Pennsylvania, Philadelphia.

White, Bruce

1994 Encounters with Spirits: Ojibwa and Dakota Theories about the French and Their Merchandise. *Ethnohistory* 41(3):369–405.

White, John

1975 Historic Contact Sites as Laboratories for the Study of Culture Change. The Conference on Historic Site Archaeology Papers 1974:9:153–163. Institute of Archeology and Anthropology, University of South Carolina, Columbia.

White, Richard

1983 *The Roots of Dependency: Subsistence, Environment, and Social Change among the Choctaws, Pawnees, and Navajos.* University of Nebraska Press, Lincoln.

1991 *The Middle Ground: Indians, Empires, and Republics in the Great Lakes Region, 1650–1815.* Cambridge University Press, Cambridge.

Wiegers, Robert P.

1985 Osage Contacts with Other Indians. In *Osage and Missouri Indian Life. Culture Change, 1675–1825* (Part I), *Final Performance Report on National Endowment for the Humanities Research Grant RS 20296,* edited by Carl H. Chapman, pp. 59–85. On file, University of Missouri, Columbia.

Willoughby, Charles C.

1903 Primitive Metal Working. *American Anthropologist* (New Series) 5:54–57.

Wilson, Curtis L., and Melville Sayre

1935 A Brief Metallographic Study of Primitive Copper Work. *American Antiquity* 1(2):109–112.

Wilson, Samuel M., and J. Daniel Rogers

1993 Historical Dynamics in the Contact Era. In *Ethnohistory and Archaeology: Approaches to Postcontact Change in the Americas,* edited by J. Daniel Rogers and Samuel M. Wilson, pp. 3–15. Plenum Press, New York.

Wilson, Samuel M., and J. Daniel Rogers (editors)

1993 *Ethnohistory and Archaeology: Approaches to Postcontact Change in the Americas.* Plenum Press, New York.

Winship, George Parker (editor)

1905 *Sailors Narratives of Voyages along the New England Coast, 1524–1624.* Houghton Mifflin, Boston.

Winters, Howard D.

1968 Value Systems and Trade Cycles of the Late Archaic in the Midwest. In *New Perspectives in Archeology,* edited by Lewis R. Binford and Sally R. Binford, pp. 175–221. Aldine, Chicago.

1981 Excavating in Museums: Notes on Mississippian Hoes and Middle Woodland Copper Gouges and Celts. In *The Research Potential of Anthropological Museum Collections,* edited by Anne-Marie Cantwell, James B. Griffin, and Nan A. Rochschild, pp. 17–34. Annals of the New York Academy of Sciences 376. The New York Academy of Sciences, New York.

Wolf, Eric

1982 *Europe and the People without History.* University of California Press, Berkeley.

Wray, Charles F., Martha L. Sempowski, and Lorraine P. Saunders

1991 *Tram and Cameron: Two Early Contact Era Seneca Sites.* Charles F. Wray Series in Seneca Archaeology, Vol. 2. Research Records No. 21. Rochester Museum and Science Center, Rochester.

Wray, Charles F., Martha L. Sempowski, Lorraine P. Saunders, and Gian Carlo Cervone

1987 *The Adams and Culbertson Sites.* Charles F. Wray Series in Seneca Archaeology, Vol. 1. Research Records No. 19. Rochester Museum and Science Center, Rochester.

Wright, Rita P.

1989 New Tracks on Ancient Frontiers: Ceramic Technology on the Indo-Iranian Borderlands. In *Archaeological Thought in America,* edited by C. C. Lamberg-Karlovsky, pp. 268–279. Cambridge University Press, Cambridge.

1993 Technological Styles: Transforming a Natural Material into a Cultural Object. In *History from Things,* edited by Steven Lubar and W. David Kingery, pp. 242–269. Smithsonian Institution Press, Washington, D.C.

Zacharias, Sandra K.

1983 Examination of 16th–17th Century Sheet Brass and Copper from Huronia. Unpublished M.A. Thesis, Department of Anthropology, University of Toronto.

Zitomersky, Joseph

1994 *French Americans-Native Americans in Eighteenth-Century French Colonial Louisiana.* Lund Studies in International History 31. Lund University Press, Lund, Sweden.

Index